HIGH PERFORMANCE ESCORTS
Mk 11 1975-1980

Compiled by
R.M. Clarke

ISBN 0 946489 63 7

Distributed by
Brooklands Book Distribution Ltd.
'Holmerise', Seven Hills Road,
Cobham, Surrey, England

CONTENTS

ACKNOWLEDGEMENTS

This is the second of two books reporting on the powerful sporting Escorts. The companion volume deals with the Mk. I machines up to 1974 and this then takes the story through the Mk. II period up to 1980.

We have in fact covered part of this area before as two years ago we published Ford RS Escorts 1968-1980. The reason for our return to the subject is that Escort owners thirst for information and have contacted us in a variety of ways and urged us firstly to search our files more carefully for different articles and secondly to be more specific by producing books on each of the model types concerned.

Our spectacular cover photograph is of Bjorn Waldegärd making good headway in chilly conditions in the 1979 Swedish Rally and we are indebted to our friends at Ford for the illustration.

We know that Escort owners will wish to join with us in thanking the management of the magazines listed below for allowing us to include their copyright and interesting stories, Asian Auto, Auto International, Autocar, Car, Cars & Car Conversions, Custom Car, Modern Motor, Motor, Motor Manual, Rally Sport, Sports Car World, What Car and Wheels.

R.M. Clarke

The wraps are off:

ESCORT RS2000!

Just before the Geneva Show Ford released pictures of the two-litre four-cylinder single ohc Escort RS2000, and quite an eyeful it looks. Using a lowered-drag nose of polyurethane flexible plastic, the 180 km/h model is 16 per cent better-streamlined than the normal Escorts, and the integral air-dam beneath the front bumper reduces front-end lift by 25 per cent. A spoiler of the same material on the rear bootlid cuts back-end lift by 60 per cent.

The plastic nose is also reckoned to be a good safety measure, and less likely to inflict serious injuries to pedestrians. The two-litre ohc Cortina engine (also used for Capri and Consul 2000, as well as the American Ford Pinto compact) is uprated by 9 kW by use of a special high-efficiency exhaust system. Peak power is 82 kW at 5500 rpm, and max. torque 162 Nm at 3700 rpm.

Ride has been improved dramatically by careful attention to spring rates, roll bar thicknesses, damper valving and so on, and the car will be factory-produced in contrast to the earlier 'Performance' Escorts which were hand-made at Ford's 'AVO' (Advanced Vehicle Operation) plant at Boreham. This has been closed, maybe only temporarily, in the light of the current climate.

So far (as I write) no more has been released about the forthcoming RS1800 with 1840 cm3 16-valve twin ohc (single, twin-choke carburettor) 93 kW engine, but it is expected to have more or less the same body as the RS2000. No prices have been mentioned for the RS2000 but presumably after the Geneva Show I shall be able to tell you.

.... and Douglas Armstrong drives

THE new Escorts are significant on more than one count. They represent an enormous company's thinking about future motoring requirements (for the immediate future at any rate), and they provide a graphic illustration of the dramatic price increases that all manufacturers will have to follow under the new production requirements of less numbers and longer-life cars.

For instance, before going off to the southern Portugal region known as the Algarve, to test the new range, a standard Escort 1100 two-door (drum brakes, rubber mats) costs in the UK some 1213 pounds (about $2171).

Now grip your chairs for a minute — the 'New Escort' 1100 two-door (with drum brakes, rubber mats etc) costs 1440 pounds ($2577 approx.) — an increase of 227 pounds (about $405)!

At the moment, Ford are way ahead of other equivalent volume-production cars, on price, and that includes marques like Fiat and Mazda besides the British makes. Obviously Ford have had to plan ahead, for there would have been nothing worse than introducing a new model at a low price, then having to bump it up within a week or two.

The other British makes will undoubtedly increase their prices before very long, so it will be interesting to see how long Ford stay in the 'expensive car' sector.

It is a fact however that Ford of Britain have for long wanted to break away from the 'T-model' image — the original high-production four-cylinder 'buggy' that put America on wheels, at minimum cost. There are those at Ford who see their products as being in the Rover/Mercedes/Volvo class. If they can maintain reasonable sales, and keep ahead on price, they will have perhaps achieved what they set out to do.

Be all that as it may, the new range of Escorts is good, and there is no doubt they will meet the motoring requirements of a great many people in this new era of high petrol prices and shortages etc. They range from 1100 cm3 to 1.6 litres, and their super-dooper Ghia models for the execs, plus four estate cars (although these are new front panels on the old cars), and the just-released RS 2000. There is also the 1840 cm3 twin ohc model yet to come.

Driving the cars in the Algarve I was most impressed by the quality of the new ride. The normal Escorts gave a near-soft ride, soaking-up the poor Portuguese country roads in the most impressive manner, yet providing good handling too. All right, on those rutty roads, if you really threw, say a 1300GL into a bend, it might patter its live back axle a bit as the lateral forces sent it outwards after the lift had 'dislodged' the tyre from a rut, but for a family car the handling must be classed as first-rate.

The Sport models and the Ghias, with lower build, slightly stiffer suspension, and thicker anti-roll bars, were really impressive on the bends and corners, but again, I was surprised by the quality of the ride.

As I mentioned recently, the live rear axle (all models) now sits on three instead of four-leaf springs, and there is an anti-roll bar which doubles as a pair of trailing arms to resist longitudinal forces. The front

MacStruts have softer coils, the torsional anti-roll bar has been beefed-up, and the damper valves have been much-modified.

First car I drove after I climbed off the BAC 111 at Faro airport was the two-door 1600 Sport in flaming red with auxiliary lights, strips, signwriting, and what else! I'm bound to say it's the sort of car that takes about 30 seconds to become familiar — the cloth-faced seats fitted like a glove, the driving position was great, and the dash layout (with matched speedo and rev-counter) was attractive and easy to read.

The 1598 cm3 pushrod ohv engine (virtually the 1.6-litre Capri motor) puts out 63 kW at 5500 rpm on a compression ratio of

Above: 1600 two-door Sport model has spirited performance and responsive handling, according to our man Armstrong.

9:1 (fuel requirement 96-97 octane), and is a really responsive unit. Two-up it would accelerate from 0-100 km/h in 13.2 sec, and would top 160 km/h in fourth.

On Michelin ZXs the roadholding was great, but lots of sandy-surfaced roads brought home the fact that the handling characteristic was understeer. In other

Below: Four-door 1300 GL shows crisp new styling to advantage. Car returned a fuel figure of 7.43 litre/km with a mild attempt at economy running.

the new Escorts in the Algarve

Above: The Ghia influence has now worked its way down the ranks to Ford's bottom-liner. This is the "Turinized" two-door 1300 version.

words, if you really wound it on with the front tyres on a sandy corner, it went straight on! On good roads of course it was terrific.

Although the 1600 Sport is, as its name suggests, a sporting Escort, it is also quiet and sophisticated. It pulls a high top gear (30.57 km/h per 1000 rpm), and at 100 km/h is very quiet indeed — in spite of a (Weber) twin-choke carburettor. Liked it.

Next morning there were 95 km/h gales coming straight in off the Atlantic (there's nothing between the Algarve and the USA!), yet the 1300 GL four-door behaved impeccably. All right, a really powerful gust would affect the steering, but it was easily coped with. There were trees down all over the place, and the poor bloody local Foresty Commission scouts were having a tough time on their pushbikes. Not to mention the local tearaways on their mobikes!

For 183 km of the second day's test course there was the option of participating in an 'economy run', sponsored by the Portuguese Mateus Rose house (makers of the 'slightly-sparkling pink wine which comes from another part of the country, and is sold in those oval-type bottles). My co-driver and I decided we'd have a half-go for we didn't want to crawl about and not discover the 1300's good or bad qualities.

So we coasted down quite a few hills, and along straights, but we kept the plot going fairly *rapido*, and after refuelling were quite pleased with our 38 mpg (7.43 lit/100 km). To give some idea of the other end of the scale, one journalist, rather fed-up with the fact that there was no index for larger engines and automatics, just blasted full-bore (two-up) around the 183 km course in a 1.6-litre automatic and averaged better than 26 mpg (10.87 lit/100 km). Which can't be bad.

The 1300 GL was incredibly quiet for a car of such modest engine capacity. The 1297 cm3 pushrod ohv motor develops 42.50 kW (DIN) at 5500 rpm on a compression ratio of 9.2:1 (demands only 97-97 octane petrol in spite of the relatively high ratio), and pulls high gearing, for a 1.3-litre, at 27.5 l km/h. Sound-deadening material has been greatly increased on the new range, and all models are notably quiet. The top Ghias are positively loaded with it, and even in 1300 form are very quiet indeed.

After a super sea-food luncheon at one of Portugal's beautiful *Pousadas* (state-owned luxury hotels — at very reasonable prices) laid-on by Mateus Rose, we set-off once again, that time in a 1300 Ghia two-door.

The seats of the Ghia models are almost in the R-R class, the finish is first-class (ought to be at 2011 pounds! about $3600!!), and the handling plus ride impressive. In spite of all the extra inbuilt equipment this little number (two-up) managed a 0-100 km/h in 13.4 sec., and hummed along at 135 km/h quite effortlessly. Whichever way one looks at it, these are very different Escorts from the ones that went before.

On the morning of departure we grabbed an 1100L four-door, and this time we were four-up with luggage. In this form (the 'L' is for 'Luxe') the smallest-engined model has cloth seats and cut-pile carpets but the drum brakes remain, although front discs can be specified at extra cost. The 'L' as we drove it costs 1586 pounds (about $2839), plenty of money, even in this day and age. There was no opportunity to test acceleration, but certainly the 1100 cruised easily at 110 km/h, and on the long straight into Faro we were able to exceed 130 km/h. The 1097 cm3 pushrod ohv motor puts out 36 kW at 5500 rpm on a compression ratio of 9:1, and pulls a 4.125:1 top gear which at 1000 rpm results in a road speed of 26.22 km/h. The extra sound-deadening plus the reasonably high-gearing of the 1100 makes it a quiet car in its class.

The quietude and comfort is emphasised in all new Escorts by the extra elbow room, and rear legroom, which has been found without resort to increasing the exterior dimensions at all. Even the boot takes 10 per cent more luggage without outside alterations.

So, there you have an outline of this most important international car — we shall now have to sit back and see if the buying public come forward with the extra money in these difficult times.

It is worthy of comment that, for instance in the UK, the Escort 1100L two-door costs 1529 pounds against the Austin Marina two-door 1.3 de luxe at 1425 pounds; the Escort 1300GL costs four-door costs 1777 pounds against the Triumph Toledo (1.2-litre) four-door at 1532 pounds); the Escort 1600 Ghia four-door at 2125 pounds compares with the 1800 Vauxhall Magnum four-door at 1842 pounds, and so on.

They're pricey at the moment but let's wait a few weeks!

ROAD TEST

FORD ESCORT SPORT 1600 2 DOOR

FOR : roomier and more refined than old Escort ; good economy for performance ; excellent seats, gearchange and instruments ; light and easy to drive

AGAINST : rather expensive ; jiggly low-speed ride ; tail slides when cornering hard

The original Escort came into being in January 1968 as a replacement for the long-running, notchback Anglia. It offered superior performance, handling and comfort, wrapped in an undeniably more attractive shell. Now, seven years later, that same car was, until very recently, Britain's second-best seller and has behind it an almost unparalleled string of competition successes. Why, then, a revamped shell now?

Like all manufacturers, Ford plan far ahead. Unlike some others they also believe in updating while sales are booming rather than using a decline in the market as a signal for improvement. They did it with the Cortina (currently Britain's best seller), they did it with the Capri. Now it's the turn of the Escort.

Like the Capri II, the new Escort is basically a reskinning job and the airy body (which inherits no panels from its predecessor) is fixed to the old platform with improved running gear. The "Kent" family of engines has been retained in the interest of economy and ease of service, the mainstream being broadened to include a 1600 cc version. Trim packs follow the latest Ford theme, running from the base model through L and GL to the more luxurious Ghia.

Our test concerns the 1600 Sport, an in-between variant that is clearly aimed at the enthusiast. Its engine is that of the now obsolete Mexico and so offers equivalent performance in a more refined car with good handling and most of the other features that distinguish the new Escort from its predecessor. This means better seating, ride and visibility (the new car has 23 per cent more glass area), improved heating, ventilation and sound insulation, better accommodation, a more comfortable driving position, and improved instruments.

At £2054 the Sport is not cheap and there are several cars that offer superior performance and comparable comfort for the price —the Opel Ascona SR and Fiat 124 ST, for example. However, few rivals blend such performance with the same measure of economy and the sort of handling that is now expected of an Escort. In the same vein, while the trim is functional rather than luxurious, the car is well equipped in a way that allows one to make optimum use of the available performance. There are powerful halogen head and spotlights, a remarkably accurate revcounter and speedometer, excellent reclining seats with built-in headrests, extra taut suspension

and low profile tyres on wide rims.

On the debit side, while the handling is good, it is not a match for Ford AVO's old Mexico and is a little too tail happy. The low-speed ride, though far better than that of AVO's now obsolete car, is a little harsh compared with that of most of the Escort's competitors.

Even so, we enjoyed the car, which endeared itself to all our drivers, even the least performance-minded ones. Though its price tag looks high just now, it no doubt allows for future increases that will affect most other makes in the near future.

PERFORMANCE

★★
★★ The 1600 cc pushrod engine is back in volume production. Though formerly fitted to Cortinas and Cap-

ris, its use for the Escort range was previously confined to FAVO's Mexico. Like the 1300, it is a stroked version of the basic 1100—all three variants sharing an over-square bore of 80.9 mm. In its current form, with emission carburetter and matching fabricated exhaust manifold, it develops 84 bhp at 5500 rpm as opposed to the 86 bhp of the obsolete Mexico. Maximum torque is still 92 lb ft at 3500 rpm.

Another change in specification is the adoption of an automatic choke which worked well: the car always started promptly from cold and warmed without fuss. With a minimal decrease in weight and increase in frontal area, the performance of the new Sport is very similar to that of the Mexico. At 97.9 mph the top speed was a little down, though a best flying quarter of 100 mph shows it has the ability to "wind

Above: Ford have submitted the design of the seats for an award. They recline fully, feature built-in neck restraints, and are very comfortable

Left: slim seat backs help to increase legroom for rear seat passengers

Right: the boot held 8.9 cu ft of our test luggage

up" in favourable conditions. However, at 11.1 s, the acceleration to 60 mph is in fact slightly better than that achieved with our early Mexico test car and matches Ford's claim exactly. In top, the 30-50 mph time takes an acceptable 10.6 s, reflecting quite good flexibility.

While this performance is more than satisfactory for a 1600, it is not particularly impressive for one in the Sport's price bracket. As our comparison chart shows, you can buy more performance elsewhere for an equivalent or even smaller outlay. In terms of drivability, however, the Sport is hard to beat. As always the Heron-headed, cross-flow engine is a willing worker. It is better insulated than before and only becomes obtrusive when revved above 6000 rpm—a pointless exercise as it is then well past peak power. We detected a hesitation in the transmission from first to second choke of the Weber carburetter and found it necessary to press hard into the carpet to get full acceleration. Though there appears to be a slight flat spot in the 30-50 mph region in top, the engine is really very tractable at low revs and will pull strongly from well under 1000 rpm. The tickover was a lumpy 700 rpm.

ECONOMY

★★★★ The formula of an efficient engine in a reasonably light, compact shell gave the Mexico a useful edge on economy over most other 1600 saloons. The Sport continues this tradition, our overall figure of 27.3 mpg comparing very well with other, some larger, cars in this class. With a computed touring consumption of 32.4 mpg, no doubt most owners will enjoy an average in excess of 30 mpg.

With its 9.0:1 compression ratio the Sport requires four-star fuel, a full tank of which will carry you some 290 miles.

TRANSMISSION

★★★★ Understandably, Ford have seen no need to change the gearbox; the single-rail shift transmission has attracted so much praise over the past few years, there was no need. However, the high-powered versions of the new Escort do benefit from a set of specially developed close ratios. The internals are also shot-peened and have high-capacity bearings.

Taking into account the higher 3.54:1 axle, these new ratios give

the Sport a slightly lower first gear, a higher second and lower third than those of the Mexico. At 18.9 mph/1000 rpm top gear is now more relaxed. The speeds available at 6500 rpm are now 36, 61 and 85 in first, second and third respectively.

The change is as light and positive as ever, though our test car did suffer from weak synchromesh on third. The clutch found on all new Escorts is the 7.5 in diaphragm variety previously used only on the 1300 cc models. It is smaller than the old Mexico's and was slightly sharper in its take-up.

HANDLING

★★★ With its light, extremely direct rack and pinion steering the Sport is a remarkably easy and responsive car to drive. Only 0.8 of a turn is needed to scribe a 50 ft circle, so many manoeuvres need little more than a flick of the wrist. This quick responsive steering will be appreciated as much when parking at the shops as when driving quickly on a twisting road.

The layout of the suspension of all the new, volume-production Escorts is the same, and is based

on the revised platform introduced in January 1974. The latest changes are confined to spring, damper and roll-bar settings and the adoption of wider and fewer leaves for the rear springs. The Sport version has extra stiff wheel rates front and rear and 175/70 series tyres on 5J sports-style steel wheels. To recap: front suspension is by MacPherson struts, coil springs, transverse links and an anti-roll bar, and the rear by live axle located by semi-elliptic springs and a combined anti-tramp and anti-roll bar.

Like the other models, the Sport's suspension has been set to give a high degree of roll stiffness and some 75 per cent reduction in understeer compared with that of previous Escorts. We found the changes in specification very noticeable; in fact we feel the car is now a little too tail happy. Certainly really bumpy surfaces can cause the back to jump out of line and it is always this end that breaks away first, sometimes rather abruptly, on wet and greasy roads. This suggests an excess of rear roll-stiffness, a theory supported by the way the inside rear wheel tends to lift and spin on tight corners.

Inevitably one tends to compare the Sport's handling with that of the Mexico, which is perhaps a bit unfair as the Mexico was a "special build" derivative with stiffened shell and superior rear-end location (it had tramp bars). Certainly the new car does not have that same fail-safe feel, though in more general terms and in comparison with all previous volume-production Escorts the handling is very good, particularly in the dry. Roll is minimal.

BRAKES

★★
★★

As the most powerful model in the new range, the 1600 Sport has servo assistance for its disc/drum brakes. Like all new Escorts it benefits from larger diameter discs than the old cars had and automatic adjustment for the rear drums. Unlike most of the small Fords we have tested in the past few years its system was totally without vice and certainly did not suffer from the pulling to one side that so many previous Escorts and Capris have done. Though quite long in its travel, the pedal was progressive and full of feel. Even our rigorous MIRA tests failed to unearth any real weakness. A pressure of only 70 lb was sufficient to send our brake meter off its scale, in excess of 1 g, and the gruelling 20-stop fade test produced a rise in required pressure of a mere 3 lb—a figure within the bounds of experimental error. The water-splash did affect efficiency briefly, but the pressure rise was small and the recovery quick.

The handbrake was well inside our requirements, producing a 0.39 g stop from 30 mph and happily holding the car on a 1 in 3 slope.

ACCOMMODATION

★★
★★

Though the previous Escort was quite roomy when it was introduced, after seven years several later models make it seem almost cramped. One of the many aims of Ford was to find more space in the new car. Their efforts are immediately apparent, in particular the extra 2in of legroom that results from the use of shallower backs for the front seats. Both head and legroom is now about as good as you can get from a package of this size. Entry into the rear is straightforward, wide doors and easily tilted front seats leaving sufficient clambering space for even the largest of adults. Once you are there, handy cutaways in the quarter panels allow plenty of elbow room.

There's no glove box in the Sport but ample oddments space is provided in the form of a full-width front parcel shelf and an adequately-lipped ledge behind the rear seat. One annoying feature of the front passenger shelf, however, was the gap in its right-hand edge that allowed pens and other such objects to roll on to the floor—and sometimes under the driver's feet.

The boot has also been enlarged and we squeezed 0.9 cu ft more luggage in than before. As before, the spare is housed upright in the nearside wing, yet the luggage area is a far more useful

MOTOR ROAD TEST No. 11/75 ● Escort Sport 1600 2 Door

GVW 594N

PERFORMANCE

CONDITIONS
Weather	Dry, sunny; wind 10-20 mph
Temperature	38-45°F
Barometer	29.9
Surface	Dry tarmac

MAXIMUM SPEEDS
	mph	kph
Banked circuit	97.9	157.2
Best ¼ mile	100.0	160.9

Terminal speeds:
at ¼ mile	74	119
at kilometer	88	141

Speed in gears (at 6500 rpm):
	mph	kph
1st	36	58
2nd	61	98
3rd	85	137

ACCELERATION FROM REST
mph	sec	kph	sec
0-30	3.3	0-40	2.5
0-40	5.3	0-60	4.8
0-50	7.9	0-80	7.7
0-60	11.1	0-100	12.0
0-70	15.5	0-120	18.3
0-80	22.9	0-140	32.2
0-90	37.7		
Stand'g ¼	18.2	Stand'g km	34.2

ACCELERATION IN TOP
mph	sec	kph	sec
20-40	9.9	40-60	5.8
30-50	10.6	60-80	6.7
40-60	10.0	80-100	6.2
50-70	11.0	100-120	6.8
60-80	14.4		

FUEL CONSUMPTION
Touring*	32.3 mpg
	8.7 litres/100 km
Overall	27.3 mpg
	10.3 litres/100 km

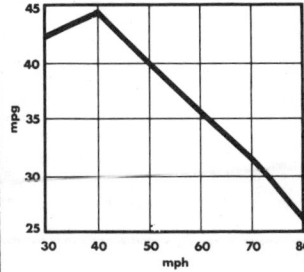

Fuel grade	97 octane (RM) 4 star rating
Tank capacity	9 galls 40.9 litres
Max range	290 miles 466 km
Test distance	1001 miles 1610 km

* Consumption midway between 30 mph and maximum less 5 per cent for acceleration.

BRAKES
Pedal pressure deceleration and stopping distance from 30 mph (48 kph)
lb	kg	g	ft	m
25	11	0.32	94	29
50	23	0.81	37	11
70	32	1.00+	30	9
Handbrake		0.39	77	23

FADE
20½g stops at 1 min intervals from speed midway between 40 mph (64 kph) and maximum (69 mph, 111 kph)
	lb	kg
Pedal force at start	35	16
Pedal force at 10th stop	38	17
Pedal force at 20th stop	38	17

STEERING
Turning circle between kerbs
	ft	m
left	29.2	8.9
right	28.75	8.8
Lock to lock	3.5 turns	
50ft ½ diam circle	0.8 turns	

CLUTCH
	in	cm
Free pedal movement	1.0	2.5
Additional to disengage	3.0	7.6
Maximum pedal load	20 lb	9.1 kg

SPEEDOMETER (mph)
Speedo	30	40	50	60	70	80
True mph	31	40	50	60	70	80

Distance recorder: 2.7 per cent slow

WEIGHT
	cwt	kg
Unladen weight*	17.0	863.5
Weight as tested	20.7	1051.0

*with fuel for approx 50 miles

Performance tests carried out by Motor's staff at the Motor Industry Research Association proving ground, Lindley.

Test Data: World copyright reserved; no reproduction in whole or in part without Editor's written permission.

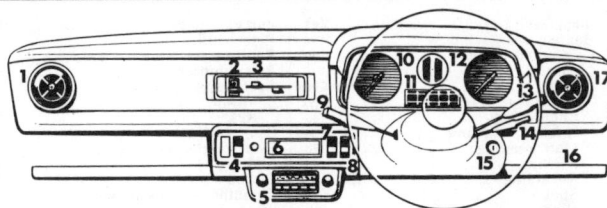

1 air vents	10	rev counter
2 fan	11	warning lights
3 heater and distribution controls	12	speedometer
	13	wipe/wash arm
4 hazard flashers	14	lights stalk
5 radio	15	ignition/ steering lock
6 ashtray		
7 heated rear window	16	parcel shelf
8 fog lamps	17	air vent
9 indicators and lamp flash		

COMPARISONS

	Capacity cc	Price £	Max mph	0-60 sec	30-50* sec	Overall mpg	Touring mpg	Length ft in		Width ft in		Weight cwt	Boot cu ft
Ford Escort 1600 Sport	1598	2054	97.9	11.1	10.6	27.3	32.4	13	0.50	5	0.50	17.0	8.9
Alfa Sud Ti	1168	1769	99.1	11.9	19.6	26.4	32.5	12	10	5	3	16.5	9.4
Colt Lancer SL 2-door	1597	1747	95.8	11.6	10.7	26.8	35.7	13	1	5	0	16.3	7.6
Datsun 180B SSS	1770	1895	104.7	10.3	9.0	24.7	31.4	13	10.75	5	2.75	20.1	9.6
Fiat 124 ST	1592	1599	99.8	10.1	13.1	25.4	29.8	13	2.75	5	4.5	18.3	10.0†
Morris Marina 1.8 TC	1798	1822	99.3	11.8	9.3	24.7	28.8	13	7	5	4	18.4	12.4
Opel Ascona SR	1897	1857	98.7	10.5	9.3	25.2	28.9	13	8	5	5	18.7	10.5
Austin Allegro HL	1748	1881	99.7	10.0	9.3	25.6	31.6	12	8.25	5	3	17.0	9.2

*in top. Fifth for Colt, Alfa, Fiat and Austin. Fourth for the others.
†measured with boxes, not cases.

MOTOR ROAD TEST No. 11/75 ● Escort Sport 1600 2 Door

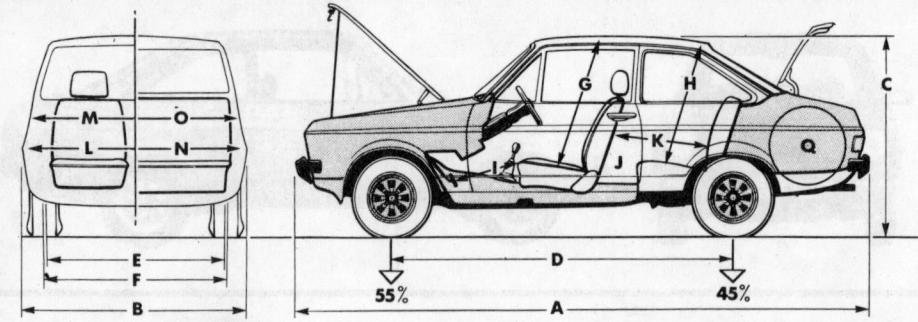

		ft	in	cm			ft	in	cm
A	overall length	13	0½	397	K	front to back			
B	overall width	5	0½	153.6		seat max	2	8	81.2
C	unladen height	4	7½	140.9		min	2	2	66.0
D	wheelbase	7	10½	240.0	L	front elbow			
E	front track	4	2	127.0		width	4	2	127.0
F	rear track	4	3	129.5	M	front shoulder			
G	com seat to					width	4	3	129.5
	roof front	3	2	96.5	N	rear elbow			
H	com seat to					width	4	6	137.1
	roof rear	3	1	93.9	O	rear shoulder			
I	pedal to seat					width	4	3	129.5
	max	1	8½	52.0	P	min ground			
	min	1	3	38.1		clearance	5½		13.9
J	kneeroom				Q	boot capacity	8.9 cu ft		
	max	1	0½	31.7					
	min		6½	16.5					

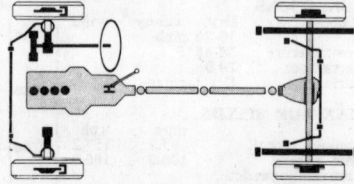

GENERAL SPECIFICATION

ENGINE
Cylinders	4 in line
Capacity	1598 cc (97.5 cu in)
Bore/stroke	80.98/77.62 mm
	(3.19/3.06 in)
Cooling	Water
Block	Cast iron
Head	Cast iron
Valves	ohv
Valve timing	
inlet opens	27° btdc
inlet closes	65° abdc
ex opens	65° bbdc
ex closes	27° atdc
Compression	9.0:1
Carburetter	Twin-choke Weber
Bearings	5 main
Fuel pump	Mechanical
Max power	84 bhp (DIN) at 5500 rpm
Max torque	92 lb ft (DIN) at 3500 rpm

TRANSMISSION
Type	4-speed manual
Clutch	7.5 in dia, sdp, diaphragm spring

Internal ratios and mph/1000 rpm
Top	1.000:1
3rd	1.418:1
2nd	1.995:1
1st	3.337:1
Rev	3.868:1
Final drive	Hypoid bevel, 3.54:1

BODY/CHASSIS
Construction	Unitary all steel

SUSPENSION
Front	Ind by MacPherson strut, coil springs, transverse link and anti-roll bar
Rear	Live axle located by semi-elliptic springs and a combined anti-roll bar and anti-tramp bracket

STEERING
Type	Rack and pinion
Assistance	None
Toe-in	0-7 mm
Camber	0° 12ft

Castor	1° 39ft
King pin	9° 7ft
Rear toe-in	0

BRAKES
Type	Front disc; rear drum
Servo	Yes
Circuit	Split
Rear valve	
Adjustment	Automatic

WHEELS
Type	Sports-style, steel
Tyres	175/70 SR-13 Michelin ZX
Pressures	

ELECTRICAL
Battery	38 amp/hr
Polarity	negative earth
Alternator	
Fuses	8
Headlights	7in dia halogen 55 watt, plus auxiliary halogen lamps on main beam

STANDARD EQUIPMENT

Adjustable steering	No	Head restraints	Yes	Parcel shelf	Yes
Anti-lock brakes	No	Heated rear window	Yes	Petrol filler lock	No
Armrests	Yes	Laminated screen	No	Radio	No
Ashtrays	Yes	Lights		Rev counter	Yes
Breakaway mirror	Yes	Boot	Yes	Seat belts	
Cigar lighter	No	Courtesy	Yes	Front	
Childproof locks	No	Engine bay	No	Rear	
Clock	No	Hazard warning	Yes	Seat recline	Yes
Coat hooks	Yes	Map reading	No	Seat height adjuster	No
Dual circuit brakes	Yes	Parking	No	Sliding roof	No
Electric windows	No	Reversing	Yes	Tinted glass	No
Energy absorb steering col	Yes	Spot	Yes	Combination wash/wipe	Yes
Fresh air ventilation	Yes	Locker	No	Wipe delay	No
Grab handles	Yes	Outside mirror	Yes	Vanity mirror	No

IN SERVICE

GUARANTEE
Duration **12,000 miles or 12 months**

MAINTENANCE
Schedule	Every 6000 miles
Free service	At 1500 miles
Labour for year (paid for)	4 hours

DO-IT-YOURSELF
Sump	5.75 pt 10W/30
Gearbox	1.5 pt SAE 80 EP
Rear axle	2 pt SAE 90 EP

Steering gear	—
Coolant	11.9 pt
Chassis lube	Nil
Contact breaker gap	0.025 in.
Spark plug gap	0.025 in.
Spark plug type	Motorcraft AGR 22
Tappets (hot)	0.010 in. inlet 0.020 in. exhaust

REPLACEMENT COSTS
(including VAT)
Brake pads (front)	£8.23
Clutch unit (exchange)	£16.87
Contact breaker set	51p
Complete exhaust system	N/A
Damper (front)	N/A
Engine (exchange)	£92.88
Front wing	£13.18
Gearbox	£36.96
Oil filter	£1.99
Starter motor (exchange)	£27.76
Windscreen	N/A

Make: Ford
Model: Escort 1600 Sport
Makers: Ford Co Ltd, Warley, Essex
Price: £1,592, plus £130 Car Tax, plus £138 VAT, equals £1,860. Extras fitted to test car: inertia seat belts, £24.86; push-button radio, £63.47; cloth trim, £14.91; reclining seats, £29.54; laminated screen, £29.54; rear fog lamps, £15.80, remote control door mirror, £15.60, giving total as tested £2,054.20.

shape than before and loading has been made easier, too, thanks to the lower rear panel.

RIDE COMFORT

★★★ With its stiff springs and low profile tyres the Sport has been aimed at the enthusiast driver. A direct result of this sporting suspension is an indifferent low-speed ride. Broken town surfaces cause the car to jigle and rock on its springs, though it rarely crashes even on the worst of bumps. As the speed increases the ride improves considerably and at high speed is quite good, certainly far better than the Mexico's. Particularly comfortable seats go a long way towards compensating for any deficiencies and we doubt if many will find even the low speed progress unacceptable.

AT THE WHEEL

★★★★ GT versions of the previous model suffered from a number of irritating faults in the control department, notably the way the fiddly rocker switches were dotted around the facia. Happily this and the other deficiencies have been attended to on all the new cars, Cortina-style stalk controls being adopted for all the major ancillaries. That to the left of the column operates the indicators, headlamp dip/flash and the horns. The larger of the two right-hand stalks operates the excellent wash/wipe system and a shorter one behind it, the lights. Much better: but still far from perfect, since the lights stalk is obscured by that for the wipers and is too far from the wheel for easy fingertip control.

We all liked the neat, leather-bound steering wheel and the pedals, though the throttle is a little too low in relation to the brake. In practice a good measure of movement on the brake brings them into line and makes for easy heel and toe changes. The seats (wrongly described as of German design in our description) brought praise from all our testers. Their excellent contours offer full support to all key areas of the anatomy and adequate fore and aft adjustment ensures comfort for tall, large men.

INSTRUMENTS

★★★★ The seats have apparently been submitted for a design award. The instruments, however, have already won aclaim, being the winning cluster first seen on the revised Cortina. The simple three-dial group is exceptionally well sited and bold in its calibrations. On the left is a large circular rev-counter marked to 7000 rpm—there is no red line. On the extreme right is a matching speedometer with trip and total mileage recorders. In the middle are the neat temperature and fuel gauges and below them a row of five warning lights. All the instruments proved to be unusually accurate, the speedometer being no more than 1 mph out at any speed and the hodometer running 2.7 per cent slow.

Left: thick-rimmed steering wheel, well-masked instruments and proper ventilation sockets distinguish the new car

Right: clever, unobtrusive air extractors in the rear pillars

Left: instruments are similar to the award-winning ones fitted in the Cortina

Right: neatly laid-out engine compartment has all service items within easy reach

VISIBILITY

★★
★★
Glass area has been increased 23 per cent on the new car and the airy effect this brings is immediately evident. The lower waistline has certainly improved visibility and though you sit quite low at the wheel the front corners of the car are clearly visible. Even though the "Coke-bottle" styling of the old car has gone, the tail-line is sufficiently upswept to make the boot invisible from the driving seat. Lateral vision is virtually uninterrupted and the main roof pillars are about as slim as they could be.

The two-speed wipers clear a good proportion of the front screen making for good wet weather visibility, and the combination of quartz halogen head and spot lights makes a nonsense of night driving, their range being more than adequate for the Sport's performance. A heated rear screen is standard and makes light work of clearing any mist on cold or damp mornings. Likewise, the defrosting action of the heater clears the front screen in seconds.

HEATING

★★
★★
Ford heaters are usually powerful and easily controlled; that of the new Escort is no exception. Two simple slide levers look after the distribution and temperature settings and a toggle switch, the powerful two-speed fan. We found the slide levers smoother running than on previous Ford test cars (this layout is standard on nearly all Ford models) and had no trouble in bringing the car up to temperature in the minimum of time. We liked the way the settings are illuminated at night.

VENTILATION

★★
★★
By our definition, the old Escort did not have proper face-level ventilation. The new eye-ball set-up is therefore a big improvement. Like those of the current Cortina, they are usually effective enough on ram pressure alone, but can be fan-boosted when required. They are easily angled to the desired position and the volume of air is simply adjusted by means of centrally controlled flaps.

NOISE

★★
★
Though the Escort does not benefit from the same degree of insulation as the GL models it does have all the deadener pads and felt of the base and L derivative plus a layer of sandwiched foam on the bulkhead. This results in a tolerably quiet car and in this respect it is certainly an improvement over the previous model. The engine remains unobtrusive until it is really revved, becoming harsh, even a little boomy from 5800 rpm onwards. There is some road noise on coarser surfaces but it never becomes objectionable. Wind noise is also low until

around the 80 mph mark, at which speed it built up quite markedly.

Though we could not detect any gearbox whine there was an intermittent vibration in the transmission when the car was lightly laden. Its disappearance at lower ride heights suggests that the differential was not running at its optimum angle.

FINISH

★★
★
For a car in this price bracket the finish is nothing special, with matt black paintwork on the facia. The optional cloth seats were neatly tailored, however, and the carpets a better quality than on some rival cars. The doors clicked shut in a pleasing fashion and the major body panels appeared to fit properly. The boot floor is sensibly trimmed in bitumised felt and the bumpers treated with a special chip-resistant epoxy paint. We felt the front indicators were rather vulnerable, being built into the base of the quarter bumpers. So are the powerful spot lights, which though sited behind the tip of the rubber overriders, are very much in line with the projecting areas of other vehicles.

FITTINGS

★★
★
Although not outwardly luxurious for a car costing well over £1800 the Sport is equipped with many sensible features that will appeal to the enthusiast Not least of these are the quartz-halogen headlights and

spotlights. A boot lamp, twin reversing, courtesy and hazard lights are also included in the specification. A sports steering wheel, sport-style road wheels, rev-counter, door mirrors, a dipping mirror, heated rear screen and 70 series tyres are also part of the pack. Inside, there are grab handles, coat hooks and armrests but no vanity mirror, clock or cigar lighter. Reclining seats with built-in headrests are standard; cloth trim is an option.

IN SERVICE

The bonnet release is sited under the facia on the driver's side. The lid itself is not self-propping. Access to the engine and its ancillaries is not as good as it used to be, the new air cleaner with temperature controlled intake partially obscuring many of the ancillaries, including the coil, distributor and oil filter. The fuse box, however, is easily located in its position on top of the bulkhead, as are the prominent washer bottle, battery and brake fluid reservoir.

The jack and wheelbrace are located together with the spare wheel in the nearside wing of the boot. The nine-gallon fuel tank is housed in the opposite corner. One key operates both door and boot locks as well as the ignition. Service intervals including oil changes for the new car are every 6000 miles, the gearbox and differential being sealed for life. The warranty is for 12 months or 12,000 miles, whichever occurs first.

RS 1800 AND 1600 SPORT

Triple C impressions of the two performance cars in the MK II Escort range

FORD'S MK II Escort range is far more comprehensive than that offered with the old model. On paper sporting types now have four basic cars to catch their eye — two Sports, at 1300 and 1600cc, and two RS models, the 1800 and 2000. In reality the choice is not quite so great: the RS1800 is only just trickling into Rallye Sport dealers and the RS2000 still seems to be a one-off show car.

It seemed right that in the issue the intrepid Andrew Dawson should reveal all about a full works RS1800 rally car that we should give you the low down on the road version — and throw in some Wislonian comments on the 1600 Sport as well. Colin in fact took the Sport all the way to Finland to watch this year's 1000 Lakes Rally (and come back, of course) while Editor Paul used the RS.

The 1600 Sport costs £1933 for the two door version and differs from lesser model Escorts in that it has the 84bhp Kent series pushrod engine of 1598cc (it's now homologated at under 1600!) and although it has the same 9.6in dia front disc brakes of the 1300 Escort it has larger, 9in x 1.7in rear drums and a servo is standard rather than an option. Spring rates front and rear are stiffer but the traditional Escort strut front/leaf spring rear is retained. At the rear there is a combined anti-roll bar and anti-tramp bracket. The gearbox has slightly higher intermediate ratios than lower power models and the final drive is 3.54, wheels being 5in wide 13in styled sports type. There are of course styling differences to put the Sport models aside from the rest of the range and equipment is pretty comprehensive, surpassed only by the luxury Ghia models. The only mechanical difference between the Sports is that the smaller engine car has a 4:125 final drive.

The RS1800 will cost its buyer £2925 or £3049 if he specifies the more lush 'Custom' trim. Mechanically the car is virtually the same as the Sport except, of course, for the engine, different rated front struts, rear radius arms instead of a roll bar and 5½in wide wheels. The bodyshell is also the Mk II equivalent of the extra-welded heavy duty Type 49 shell of the Mk I.

The RS engine is now all alloy and 1840cc. The sixteen valve, dual overhead cam, design is retained but the carburation is one progressive dual choke Weber instead of a pair of 40 DCOE's. Power is 115bhp at 6000rpm. Visually the RS stands out with the aid of a front air dam and boot spoiler.

Above: the RS1800 interior. Below left the bespoilered 1800, right the 1600 Sport

Wislon's trip to Finland was really quite a marathon — 3,300 miles through Europe and back to look at the 1000 Lakes Rally. Average fuel consumption for the trip was a very creditable 33.63mpg and Colin and Fred Henderson used only one and a half pints of oil. Colin reports that most of the motoring was done at 70mph-plus speeds.

Looking back on the trip Colin and Fred remember the 1600 Sport as a very impressive car. It had more interior room than the Mk I Escorts both of them have owned and on several occasions took four people on long trips with ease. Wind noise was low and the only criticisms were of a whining rear axle (which never got any worse) and slight electrical failures which caused a headlamp and spotlamp to need attention. As far as ride is concerned Colin rates the Sport as a very comfortable car with predictable handling.

The RS1800 is basically the same as the Sport when it comes to general comfort and handling. The plush front seats (reclining buckets similar to those fitted to the Mk I RS2000) mean that rear leg room is restricted but the driving position is very good.

The big thing about the RS is, of course, the engine. And that's the part of the car that makes our Editor wonder if the RS1800 is really worth the extra over the 1600 Sport. Yes, it's a super specification engine, but in normal road use the car felt very little faster than the Sport. At the top end of the rev range the RS motor seemed flat, probably due to being under carburated, and it was most definitely noisier than the pushrod Kent unit. The higher gearing meant more relaxed cruising but otherwise there's very little in the specification to warrant the RS instead of the cheaper car. However, the 1975 RS is most definitely a far more civilised vehicle than it was during the days of the Mk I Escort.

It's difficult to see exactly who will buy the RS1800 when they manage to get hold of one. For normal motoring it would not really seem to be worthwhile and for the man wanting to go rallying the logical way to prepare a car is to buy a bodyshell and build in the appropriate parts.

The 1600 Sport is a very nice package, the price is fair and the Mk II Escort concept is way in advance of the old model. The RS 1800 is, once again, a nice car but for normal road transport it seems hard to be able to justify that extra £1000. Perhaps the RS2000 — when it appears — will be the ideal compromise. ■

AUTO TEST

Ford Escort 1·3 Ghia

1,297 c.c.

Up-market version of Ford's new Escort shows significant gains in many areas, especially quietness and handling. Performance and economy both better than average, standard of equipment high. Good brakes, excellent gearbox, generally very easy to drive. Ride no better than average, but seat comfort good

Escort shape does not appear very aerodynamic, with a bluff grille encompassing the rectangular halogen headlamps which are a Ghia feature. Note the bumper with its inset rubber strip (and sidelamps) and overriders. Flared wheel arches ensure good wheel clearance

FORD waited seven years to replace the old Escort, and when they did the new model was so close to the old one in basic specification as to make the casual observer wonder why they bothered. There is more to it than that, of course. A firm like Ford does not spend millions of pounds re-tooling to produce something pointless. The answer lies in the word *basic*. The new Escort may be much the same size and weight as its predecessor, may use the same engines and transmissions, the same suspension arrangement; but the changes which have been made – and there are many – are all aimed at eliminating the weaker points and improving on the old car.

The Escort 1·3 Ghia, the subject of this test, is a top-of-the-line model powered by the twin-carburettor 1300GT engine and distinguished by a trim package developed by Ghia of Turin. Ghia is now entirely owned and run by Ford and devotes its energies to styling such packages for Ford models – the Mustang, Granada, Capri and now the Escort. In other respects the car resembles the 1·3 Sport (there being no GT model as

such). The Ghia package as applied to the Escort includes special carpeting, cloth trim for seats and door panels, a wood-veneer instrument panel and tinted glass.

Performance and economy

In this twin-carburettor, high-compression form the 1·3-litre Ford engine produces 70 bhp at a peak of 5,500 rpm, a handsome advance on the 57 bhp of the single-carburettor engine fitted to the L and GL versions of the new model. There is little improvement in torque, however, though the torque peak goes up from 3,000 to 4,000 rpm which must tend to make the engine less flexible. In recognition of this, the final drive is lowered from 3·89 to 4·125 (the same as for the 1100 model). This gives overall gearing of 15·9 mph per 1,000 rpm on 155–13in. tyres: apparently rather low by the standards of the class. Another change compared with the down-market 1300s is the adoption of the closer-ratio GT gearbox in which third gear is slightly higher, and second and first gears considerably higher.

The test car's maximum speed underlines our question mark against the choice of

gearing. At a mean 93 mph the Ghia is fast by 1300 standards and well up to the performance of the prestige 1½-litre cars such as the Vanden Plas 1500 and the Triumph 1500TC; but 93 mph takes it to 5,850 rpm which is well over its peak, whereas the 3·89 final drive would mean this speed corresponded almost exactly with peak power.

This is not to say the Ghia feels noisy or strained at maximum speed. Far from it: careful engine mounting and the Ghia sound-deadening package have done their work well and it is possible to converse with only slightly raised voices at 90 mph. But it is a typically German approach slightly to under-gear a car in the interests of speed-stability on *autobahn* gradients, and one wonders if the choice of final drive was Cologne's rather than Dagenham's.

Lower gearing normally shows benefits in acceleration, but at least partly outweighed by the need to change up earlier. With the close-ratio box, the Ghia's ratios are nicely spaced and judged to give it maxima of just over 30, 50 and 70 mph in first, second and third. In practice it works out very well, with

GVW 607N

AUTOTEST

Ford Escort 1·3 Ghia

no obvious gaps between the gears, and top is pulling lustily by the time one changes up from third.

As in maximum speed, so the acceleration times are excellent for the class and the

equal of many bigger-engined cars. Despite the simple rear suspension layout there is no sign of axle tramp during full-throttle standing starts, but one has to be brutal letting in the clutch if the wheels are to spin

sufficiently to keep the engine "on the cam" away from the line. It pays to hang on to each gear beyond 6,000 rpm if looking for ultimate performance; 60–70 mph takes 5·5sec in third, for example, com-

pared with 7·4sec in top. Up to this point there is no sign of the engine running out of breath, but valve bounce intercedes quite suddenly at 6,600 rpm. There is no red line on the rev counter, which reads to 8,000 rpm. In our test car the rev counter was virtually accurate throughout its range, whereas the speedometer under-read by 3 mph at maximum

Comparisons

MAXIMUM SPEED MPH

Renault 12TS	(£1,667)	94
Escort 1·3 Ghia	(£2,011)	93
Hillman Avenger 1600GL	(£1,626)	92
Triumph 1500TC	(£1,791)	91
Vanden Plas 1500	(£2,076)	90

0–60 MPH, SEC

Renault 12TS	12·9
Triumph 1500TC	13·2
Escort 1·3 Ghia	13·5
Hillman Avenger 1600GL	14·5
Vanden Plas 1500	16·7

STANDING ¼-MILE, SEC

Renault 12TS	19·3
Triumph 1500TC	19·4
Hillman Avenger 1600GL	19·8
Escort 1·3 Ghia	19·9
Vanden Plas 1500	20·9

OVERALL MPG

Renault 12TS	30·0
Escort 1·3 Ghia	29·3
Vanden Plas 1500	27·2
Triumph 1500TC	27·0
Hillman Avenger 1600GL	25·0

Performance

ACCELERATION SECONDS

True speed mph	Time in Secs	Car Speedo mph
30	4·0	29
40	6·4	38
50	9·0	48
60	13·5	58
70	18·6	68
80	29·9	77
90	—	87

Standing ¼-mile
19·9sec 72 mph

Standing kilometre
36·3sec 84 mph

Mileage recorder:
3 per cent under-reading

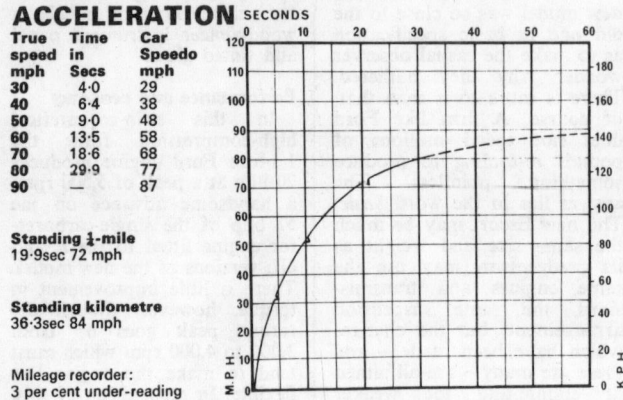

GEAR RATIOS AND TIME IN SEC

mph	Top (4·125)	3rd (5·85)	2nd (8·23)
10–30	—	8·0	5·1
20–40	11·8	7·3	4·8
30–50	11·4	7·1	5·2
40–60	12·3	7·7	—
50–70	14·0	9·7	—
60–80	17·8		

GEARING
(with 155–13in. tyres)

Top	15·9 mph per 1,000 rpm
3rd	11·2 mph per 1,000 rpm
2nd	8·0 mph per 1,000 rpm
1st	4·8 mph per 1,000 rpm

MAXIMUM SPEEDS

Gear	mph	kph	rpm
Top (mean)	93	150	5,850
(best)	96	155	6,040
3rd	73	118	6,500
2nd	52	84	6,500
1st	31	50	6,500

BRAKES

Fade (from 70 mph in neutral)
Pedal load for 0·5g stops in lb

1	35		6	30–35
2	30–35		7	30–40
3	30–35		8	30–40
4	35–40		9	30–40
5	35–40		10	30–45

RESPONSE (from 30 mph in neutral)

Load	g	Distance
20lb	0·20	151ft
40lb	0·57	53ft
60lb	0·80	38ft
80lb	0·94	32ft
100lb	1·00	30ft
Handbrake	0·30	100ft
Max Gradient	1 in 4	

CLUTCH

Pedal 21lb and 5½in.

Consumption

FUEL
(At constant speed – mpg)

30 mph	56·4
40 mph	51·4
50 mph	45·0
60 mph	39·2
70 mph	33·1
80 mph	27·9
90 mph	21·1

Typical mpg 32 (8·8 litres/100km)
Calculated (DIN) mpg 30·1
      (9·4 litres/100km)
Overall mpg 29·3 (9·6 litres/100km)
Grade of fuel Premium, 4-star
      (min 97RM)

OIL
Consumption (SAE 10W-40) Negligible

TEST CONDITIONS:
Weather: Fair
Wind: 15 mph
Temperature: 5 deg C (40 deg F)
Barometer: 29·9in. Hg
Humidity: 85 per cent
Surface: Dry concrete and asphalt
Test distance 752 miles

Figures taken by our own staff at the Motor Industry Research Association proving ground at Nuneaton.

Dimensions

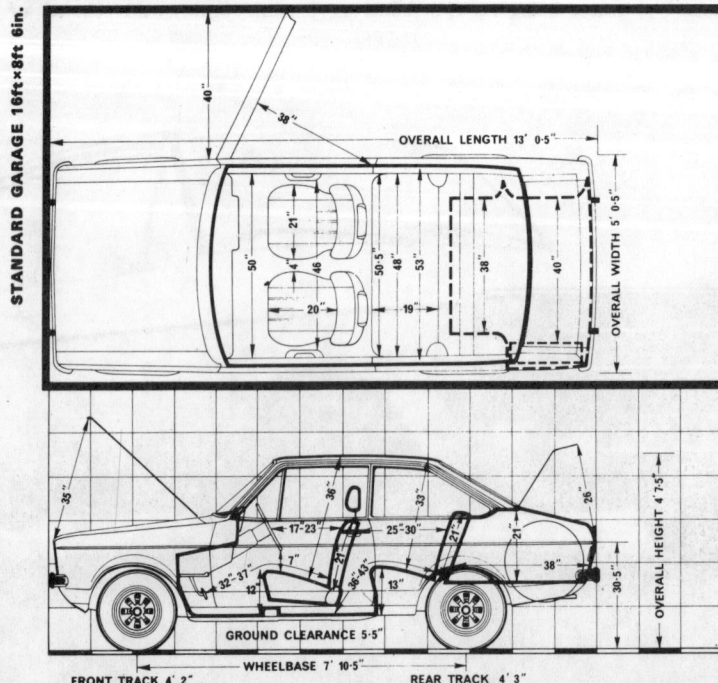

STANDARD GARAGE 16ft x 8ft 6in.

OVERALL LENGTH 13' 0·5"
OVERALL WIDTH 5' 0·5"
OVERALL HEIGHT 4'·75"
GROUND CLEARANCE 5·5"
WHEELBASE 7' 10·5"
FRONT TRACK 4' 2"
REAR TRACK 4' 3"

TURNING CIRCLES:
Between kerbs
L, 31ft 2in. ; R, 30ft.6in.
Between walls
L, 33ft 0in. ; R, 32ft 4in.
Steering wheel turns, lock to lock 3·6

WEIGHT:
Kerb Weight 17·6cwt (1,966lb–892kg) (with oil, water and half full fuel tank)
Distribution, per cent F, 53·4; R, 46·6
Laden as tested: 21·7cwt (2,436lb–1,104kg)

speed, and the mileage recorder was 3 per cent pessimistic.

As its power output indicates, the 1300GT engine is an efficient little power unit and promises good economy given a car of reasonable weight and gearing. In the Escort it gives better than average results. The steady-speed consumption graph falls almost in a straight line from 56·4 mpg at a steady 30 mph, to 21·1 mpg at 90 mph. Between those extremes, it is up to the driver to make his own trade-off of speed against economy, since there is no point at which the consumption suddenly gets worse. The key figure to remember, perhaps, is 39·2 mpg at a steady 60 mph.

Our own brim-to-brim fuel consumption varied from 27·1 mpg for the period which included the testing at MIRA, to 31·7 mpg for a gentler drive from the test track to the south coast and then into London. The touring consumption calculated on the DIN formula is 30·1 mpg, but we would expect most owners to manage 33–35 mpg on the basis of our own results. Oil consumption during our test period was too small to measure.

Handling and brakes

There was nothing much wrong with the old Escort's handling except that the back end got a bit untidy towards the limit, while the car felt likely to (but rarely did) roll if a clumsy driver got it really sideways. Since the new car

Specification
Ford Escort 1·3 Ghia

FRONT ENGINE, REAR-WHEEL DRIVE

ENGINE
Cylinders	4, in line
Main bearings	5
Cooling system	Water; pump, fan and thermostat
Bore	81·0mm (3·19in.)
Stroke	63·0mm (2·48in.)
Displacement	1,297 c.c. (79·2 cu. in.)
Valve gear	Overhead: pushrods and rockers
Compression ratio	9·2 to 1. Min octane rating: 97RM
Carburettor	Weber twin-choke
Fuel pump	Mechanical
Oil filter	Full-flow, replaceable cartridge
Max power	70 bhp (DIN) at 5,500 rpm
Max torque	68 lb. ft. (DIN) at 4,000 rpm

TRANSMISSION
Clutch	Diaphragm-spring, 7·5in. diameter
Gearbox	4-speed, all-synchromesh
Gear ratios	Top 1·0
	Third 1·42
	Second 1·99
	First 3·34
	Reverse 3·87
Final drive	Hypoid bevel, ratio 4·125-to-1
Mph at 1,000 rpm in top gear	15·9

CHASSIS and BODY
Construction	Unitary, with steel body

SUSPENSION
Front	Independent: MacPherson struts, lower links, coil springs, telescopic dampers, anti-roll bar
Rear	Live axle, semi-elliptic leaf springs, telescopic dampers, anti-roll bar

STEERING
Type	Rack and pinion
Wheel dia	15·0in.

BRAKES
Make and type	Disc front, drum rear, split-circuit
Servo	Vacuum type
Dimensions	F 9·6in. dia
	R 9·0in. dia, 1·7in. wide shoes
Swept area	F 194·3 sq. in., R 94·7 sq. in.
	Total 289·0 sq. in. (266 sq. in./ton laden)

WHEELS
Type	Sports-styled steel disc, 4-stud fixing, 5in. wide rim
Tyres – make	Michelin ZX
– type	Radial ply tubed
– size	155–13in.

EQUIPMENT
Battery	12 Volt 38 Ah.
Alternator	35 amp a.c.
Headlamps	Rectangular halogen, 110/110 watt (total)
Reversing lamp	Standard

Electric fuses	8
Screen wipers	2-speed
Screen washer	Standard, electric
Interior heater	Standard, air-blending type
Heated backlight	Standard
Safety belts	Extra
Interior trim	Cloth seats, pvc headlining
Floor covering	Carpet
Jack	Screw pillar type
Jacking points	2 each side under sills
Windscreen	Tinted toughened (laminated extra)
Underbody protection	Zinc-phosphate treatment prior to painting, bitumastic under wheel arches

MAINTENANCE
Fuel tank	9 Imp gallons (41 litres)
Cooling system	8·8 pints (inc heater)
Engine sump	5·7 pints (3·25 litres) SAE 10W/30. Change oil every 6,000 miles. Change filter every 6,000 miles.
Gearbox	1·6 pints. SAE 80EP. Check every 6,000 miles
Final drive	2·0 pints. SAE 90EP. Check every 6,000 miles
Grease	No points
Valve clearance	Inlet 0·008in. (cold). Exhaust 0·022in. (cold)
Contact breaker	0·025in. gap; 39deg dwell
Ignition timing	6deg BTDC (static)
Spark plug	Type: Motorcraft AGR 22. Gap 0·028in.
Tyre pressures	F 22; R 24 psi (normal driving) F 24; R 36 psi (full load)
Max payload	970lb (440kg)

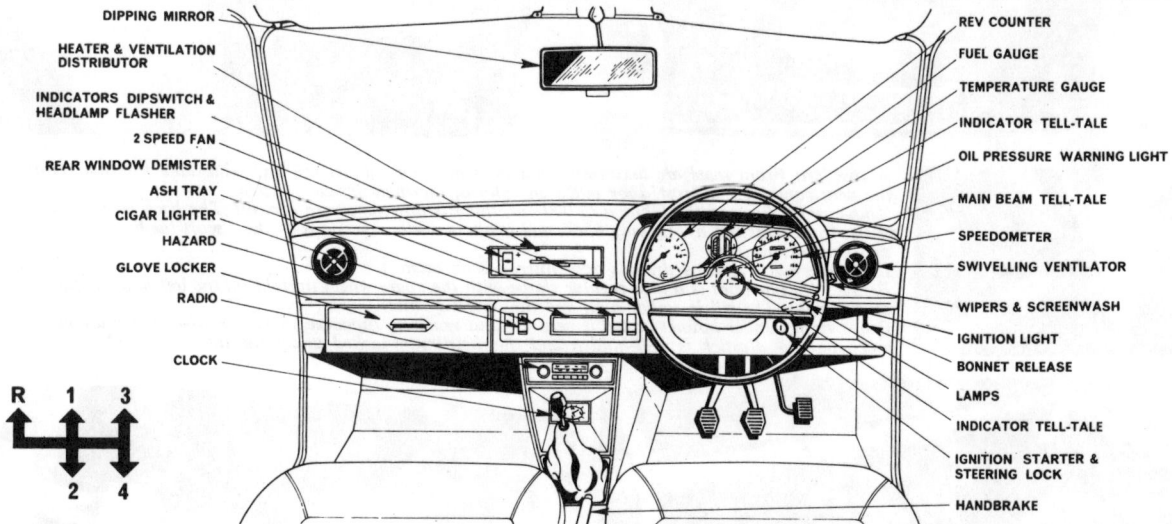

DIPPING MIRROR
HEATER & VENTILATION DISTRIBUTOR
INDICATORS DIPSWITCH & HEADLAMP FLASHER
2 SPEED FAN
REAR WINDOW DEMISTER
ASH TRAY
CIGAR LIGHTER
HAZARD
GLOVE LOCKER
RADIO
CLOCK

REV COUNTER
FUEL GAUGE
TEMPERATURE GAUGE
INDICATOR TELL-TALE
OIL PRESSURE WARNING LIGHT
MAIN BEAM TELL-TALE
SPEEDOMETER
SWIVELLING VENTILATOR
WIPERS & SCREENWASH
IGNITION LIGHT
BONNET RELEASE
LAMPS
INDICATOR TELL-TALE
IGNITION STARTER & STEERING LOCK
HANDBRAKE

Servicing

	6,000 miles	12,000 miles	18,000 miles
Time Allowed (hours)	3·0	3·0	3·1
Cost at £4.30 per hour	£12.90	£12.90	£13.33
Engine oil	£1.73	£1.73	£1.73
Oil Filter	£1.99	£1.99	£1.99
Air Filter	—	—	£1.76
Contact Breaker Points	£0.51	£0.51	£0.51
Sparking Plugs	£2.07	£2.07	£2.07
Total Cost:	£19.20	£19.20	£21.49

Routine Replacements:	Time hours	Labour	Spares	TOTAL
Brake Pads – Front (2 wheels)	0·36	£2.58	£8.23	£10.81
Brake Shoes – Rear (2 wheels)	1·12	£5.16	£9.39	£14.55
Exhaust System	0·42	£3.01	£14.23*	£17.24
Clutch (centre+driven plate)	2·18	£9.89	£16.87	£26.76
Dampers – Front (pair)	1·30	£6.45	£33.82*	£40.27
Dampers – Rear (pair)	1·48	£7.74	£13.47*	£21.21
Replace Drive Shaft	0·30	£2.15	£26.82*	£28.97
Replace Alternator	0·30	£2.15	£31.17	£33.32
Replace Starter	0·30	£2.15	£27.76	£29.91

*These prices not yet fixed for new Escort. Prices given are last available for old Escort plus 50 per cent allowance for inflation.

AUTOTEST

Ford Escort 1·3 Ghia

has essentially the same suspension its cornering behaviour does not greatly differ, but the handling has been improved towards the limit and made a good deal more forgiving.

The steering, by rack and pinion as before, is beautifully light to the point of disconcerting a driver who comes to it from a car with heavier controls. Response is crisp and instant. There is some fightback on bad surfaces, but hardly more than is needed to make the driver aware what is going on underneath him. The gearing – 3·6 turns from lock to lock – is well-judged considering the compact 31ft turning circle.

The precision of the steering gives the impression that the Escort's handling is neutral, but in fact the car understeers very consistently. The degree

of understeer increases the harder it is cornered, but can be partially overcome with the use of full power. Lifting off the accelerator in mid-corner produces a mild tucking-in of the nose as the car slows down. The sheer consistency of the handling enables the driver to exploit its limits in the safest possible way, while the accuracy with which speed can be controlled independently of the cornering line gives a feeling of being able to go deep into a corner without risking an irretrievable situation.

Any light car with accurate steering can of course be set up for a tight corner by "tweaking" the tail outwards, and in the dry the Escort will oblige smoothly and tidily. If the movement is overdone, the excess speed is rapidly

scrubbed off; the old model's feeling of being about to topple over has vanished. In the wet, the test car demanded a good deal more skill for the same technique to succeed, for the Michelin ZX tyres enabled the back end to slide rather more quickly than we would have expected. Indeed, while the roadholding on dry surfaces was beyond reproach, driving in the wet called for a lot more delicacy.

For the size and weight of car, the Escort's brakes are big and it would have been surprising had they turned in anything but a good performance. The pedal effort is well judged, and there is no sign of over-servoed snatchiness; a pressure of 60lb gave about a 0·8g stop, after which the curve steepened so that 100lb was needed for the best stop of

1g indicated, with the front wheels locked, the car slewing slightly left. In the fade test, the pressure needed towards the end of each stop increased at first, then stabilized as the brakes warmed up, after which the test was completed without any sign of distress.

The handbrake proved well able to lock the back wheels when used alone on the level, giving a 0·3g emergency stop. It also held the car either way on the 1-in-3 test hill, but the car could not manage a restart on this gradient; it tackled the 1-in-4 with ease.

Comfort and convenience

When we first had a chance to drive the new Escort, we felt the ride had improved compared with the older car. That may be so, but the ride is still no better than average for the class and it is in this area that the effect of the simple suspension layout is most keenly felt. Ford have done their best by increasing wheel travel and changing

Above left: Front seats are luxurious, well-shaped, cloth-upholstered and adjustable for rake. Note the neat combined armrest/door pulls, an echo of German design practice
Above right: Even in the Ghia, a £2,000 car, there is no rear centre armrest. The back seat is cramped if the front seats are a long way back, and in the two-door versions of the Escort (as here) entry and exit are not easy
Left: Internally-adjusted driver's door mirror is an extra
Below left: The boot is a nice regular shape with the spare wheel upright on the left and the fuel tank on the right. The sill is rather high and the opening small
Below right: Under-bonnet layout is neat without reaching Japanese standards. Most items are easy to reach but the dipstick is well-hidden and the distributor tucked under the inlet manifold

19

spring rates, but the Escort ride only feels good on smooth surfaces. Along the typical British secondary road it becomes restless, increasingly so the faster it is going, and on bad surfaces some pitching is evident. Single potholes cause a crash of complaint from the structure, but with the help of a little braking over the crest, the Escort takes hump-backed bridges very well.

If the driver is uncomfortable, the simple answer is to slow down, since the low-speed ride is well controlled with no sign of the harshness that afflicts some cars with a good high-speed ride. The driver should remember, though, that the ride in the back is worse than in the front seats – against which should be balanced the fact that (as is often the case in Ford cars) the ride improves noticeably when the car is fully laden.

The Ghia front seats are one of the better features of the Escort in this form. They are fitted with head restraints and they recline, but more to the point they have very good cloth upholstery and are well shaped to support their occupants. The range of fore-and-aft movement is sufficient for drivers from 5ft 2in. to 6ft 2in., though the latter tend to have the seat all the way back. The driving position is well laid out, even though some keen drivers might prefer the wheel set a little lower. As in all Fords, the pedals are well spaced and operate through natural and sensible arcs.

The controls are indeed one of the best points of the car. The clutch, gearchange and accelerator linkage all back up the steering and brakes with the same feeling of lightness and precision, making the Escort a very easy car to drive smoothly.

Ford have concentrated their minor controls into three column-mounted stalks, which in the face of it is a very good idea. Sadly, the detail execution does not quite live up to the promise, and most of our drivers found themselves confusing one stalk with another on occasion. When analysed, their criticisms were basically three. First, the lights and wiper stalks operate in an unnatural (to the British) sense: up for on. Second, the lights stalk, though behind the washer stalk, is the shorter of the two and therefore not easy to find in a hurry. Third, the wiper stalk was several times confused with the indicator stalk on the opposite side of the column, with drivers switching on the wipers instead of indicating a left turn.

The heater controls are illuminated – a good point – and the heater offers excellent output though in the test car

its temperature control was not as progressive as we would have liked. Distribution however was good, with extremely effective demisting action. A heated rear window is standard on all Escorts other than the base 1100 model. Ventilation, via the familiar "eyeball" vents at either end of the facia, proved entirely adequate in winter conditions; it remains to be seen if the mass flow is sufficient to cope with a hot and humid summer.

The noise level is one area where Ford have effected an improvement which, compared with the old Escort, might be described as dramatic. This at least is the impression gained in the Ghia which admittedly has the advantage of an extra sound-deadening package. Even so, the absence of mechanical noise or body boom is most notable, while road noise only makes its presence felt on gravel-dressed surfaces (or the grooved sections of M1 and M40). Wind noise presents more of a prob-

lem at high speed, and probably makes the greatest single contribution to the overall level in the 70–90 mph speed band. At 70 mph the car is quiet enough for subdued conversation, and the radio hardly needs to be turned up from its static setting.

With a flatter roof, deeper windows and slimmer rear quarter pillars than the previous Escort, the new car offers much better visibility for the driver and can be placed accurately and with confidence in a traffic stream. The extreme tail, however, cannot be seen when reversing. Two-speed wipers and electric washers are standard across the Escort range, and the

wipers sweep a good arc of the screen. Halogen headlamps are standard in the Ghia, but were not as effective as we anticipated. They have a good spread of light, but their range was limited and dipped beam was a sad contrast. A dipping mirror and reversing lights are standard; so is a four-way hazard warning flasher.

Since the new Escort uses the same wheelbase as its predecessor, there is no point in looking for significantly improved interior space, and one immediately becomes aware of this in the back. The back seat is wider than before, but head and knee room are minimal if two large occupants are installed in the front, and the back seat itself is too upright for comfort on a long journey. In the two-door car (such as was our test car) entry is far from easy since it is only the back of the front seat, rather than the whole seat, that tips forward to provide access to the rear.

Escort is practically a fastback design in the way its rear window slopes to meet the boot lid. Vinyl roof is a Ghia feature. Rear lamp clusters are neat; rearguard fog lamp is an extra

Living with the Escort Ghia

Oddly, perhaps, one cannot have automatic transmission with the 1·3-litre Escort Ghia because Ford do not offer their C3 box with the 1300GT engine; to have a Ghia automatic you must take the 1·6-litre version.

Interior stowage for oddments is sadly limited. A Ghia feature is that the open parcels shelf of lesser models is given a lid to turn it into a glovebox, but it is low down and comparatively small. The boot on the other hand is a useful shape and size, marred only by the high sill over which

luggage must be lifted. The spare wheel is installed upright on the left of the boot, matched by the fuel tank strapped in the right hand rear wing.

The tank holds 9 gallons, sufficient for a safe range of 250 miles which can only just be considered satisfactory. The filler, with a flush-fitting cap in the right rear wing, takes the last gallon much more willingly than is the case with many of today's cars.

The bonnet is released from inside the car and needs propping in the open position. The engine compartment looks tidy and some items are easily reached, but the dipstick is ridiculously concealed beneath the pancake air filter and the distributor would be far from easy to work on *in situ*. On the service side, the Escort makes no more demands than the previous model. The service interval is 6,000 miles, at which time the oil and oil filter are replaced; the transmission oil does not require replacement, and there is no chassis greasing.

In conclusion

As we said at the beginning, it is not obvious on paper why Ford should have replaced the old Escort with this one. On the road it is obvious at once. The new car is altogether nicer to drive, and one can point immediately to improvements in handling, noise level and driver visibility to justify the change. With the slight reservations we have expressed, the interior layout is much better too. A year ago we would have reckoned the car too small, for the back seat is cramped just as its predecessor's was. But that has enabled Ford to hold down the overall size and weight, with the result that the Escort is above average on performance and economy, and in this form is a match for some cars with significantly bigger engines. With the new Escort, in fact, Ford have renewed their challenge in this vital section of the market and proved that the potential of the simple, conventional but carefully engineered car is not yet exhausted. □

MANUFACTURER:
Ford Motor Co. Ltd., Warley, Brentwood, Essex

PRICES		EXTRAS (inc VAT)	
Basic	£1,721.00	Automatic transmission	£145.67
Special Car Tax	£141.42	Push-button radio*	£63.47
VAT	£148.99	Remote-control door mirror*	£15.60
Total (in GB)	**£2,011.41**	Rear fog warning lamp	£15.80
Seat Belts (inertia-reel)	£24.86	Laminated windscreen	£29.54
Licence	£25.00	*Fitted to test car	
Delivery charge	N/A		
Number plates	£6.50	Insurance	Group 4
Total on the Road (exc insurance)	**£2,067.77**	**TOTAL AS TESTED ON THE ROAD**	**£2,146.84**

Here is Ford's answer for the
Bathurst two-litre honors stakes

THE NEW ESCORT RS1800

And we could see it here before the end of the year!

LOOK OUT TWO-LITRE Bathurst racers, there's a new contender on the horizon. Along with its new range of Escort IIs Ford of Britain has released a more punchy Escort RS, with a 16-valve 1840 cm^3 motor . . . and we could see it here before the end of the year.

Ford candidly admits the car has been designed as the basis for a class-winning contender in motor sport and the new top-of-the-line Escort is a direct descendant of the highly successful RS1600.

Though Ford Australia is no longer directly involved in motor sport the company still takes a big interest in the fortunes of Ford privateers and is not above helping them out by making good equipment available.

With more emphasis on the smaller classes in racing — witness Amaroo Park's decision to limit its $15,000 series to three-litre cars —

because of a swing in public preferences to smaller cars the new Escort RS would be an excellent vehicle with which Ford could capture considerable publicity for its economy range.

To date the ageing 1600 cm^3 twin-cam Escort GT has kept the Ford flag flying in the under-two-litre class, but drivers are having an increasingly tough time coping with later model opposition. With a new car and the more sophisticated 1840 cm^3 production BDA engine the company's class-winning fortunes would be revitalised to the stage where the Escort is once again the most favored under two-litre car.

There is a chance that RS Escorts destined for Australia could be fitted with a more powerful two-litre motor. This car will be the successor to the Escort RS2000 which was produced in Britain and

Germany from 1973 onward and uses a specially developed version of the two-litre sohc motor fitted to the Cortina.

The new RS Escort was a significant factor in Ford's decision to expand its European rally program this year and this could lead to a return of Ford participation, direct or indirect, in events like the Total Oil Southern Cross. This rally's increased status, and the wide publicity gained by Japanese manufacturers in winning it, have again aroused Ford's interest and it is known that several executives have been casting around for a car which could be competitive for the event.

To allow for mass production the new Escort RS has been designed as simply as possible and uses a heavy-duty version of the standard shell. Suspension layout is basically the same as on the present

Escort though attention has been paid to refining the ride and improving the handling. At the front the car has been fitted with a thicker anti-roll bar and more progressive bump stops. At the rear a new anti-roll bar, and wider springs with three instead of four leaves are on the live axle.

The production RS1800 is fitted with a single twin-choke carburettor (instead of the twin Webers on the RS1600) to improve economy, and it develops 93 kW (125 bhp) at 6500 rpm. European compeition versions will have the 1840 cm^3 engine stretched to two litres with power outputs of 171 kW (230 bhp) for rally cars and 205 kW (275 bhp) for racing cars. The competition models will also have five-speed gearboxes, triple plate clutches and limited slip diffs. Externally the competition models will differ in the adoption of front and rear spoilers, which are primarily designed for circuit racing, and revised wheel arch extensions to cover the optional 177 mm (7 in.) wide wheels.

The full-house` competition Escort will not be seen in Australia unless Ford of England supplies a car for an event like the Southern Cross but the production Escort RS, in 1.8 or two-litre form, is likely to be landed in Australia in limited numbers before the end of the year and will give small-capacity Ford racers renewed muscle with which to face the competition from Alfa Romeo, Mazda and Leyland's Triumph Dolomite Sprint.   *

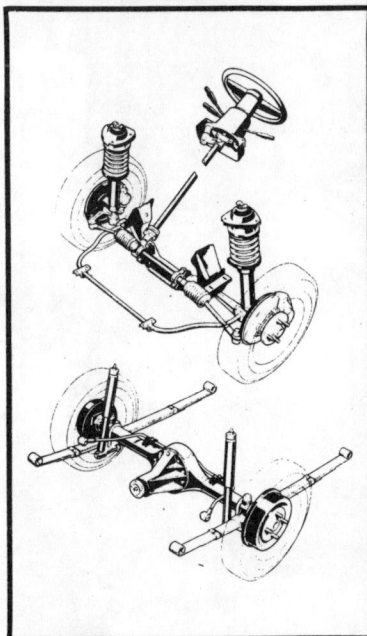

Interior has more comfortable, cloth trimmed seats and small padded steering wheel. Instrument cluster has been redesigned.

Top:
Sporty models of the range come with black bumpers front and rear, twin black door mirrors plus quartz halogen driving and headlights.

Far Left:
Brand new body, increased visibility are features of the entire new Escort range. This is the 1600 Sport, which will probably head the line-up readily available in Australia. Topline RS1800 model shares same body, mini fender arches and low ride height but is powered by 16-valve 1840 cm^3 version of old RS1600 BDA engine.

Left:
Suspension of new Escorts remains basically the same as before, though front and rear anti-roll bars are new and rear springs are wider, but with only three leaves instead of four. Multi-function stalks on steering column take the place of most dashboard switches.

12,000 MILES ON
FORD ESCORT 1600 SPORT

Despite the similar specification and sporting slant, Ford's 1600 Sport is a very different animal from the old Mexico. It still has stripes and a sporty steering wheel, but the sophistication level has gone up with the price.

Let's be honest. When I learned that my long term Renault 5 was to be replaced by Ford's new Escort – and a 1600 Sport, no less – I was excited.

Not only was I due for more contented cruising on my regular weekend sorties up and down the length of the M1, but it was going to be a fascinating exercise comparing the new sporty Escort with its predecessor, the very successful Mexico.

During 1972 I had put many road and rally miles on one of these cars (which itself succeeded the Mini Cooper as a good value, easy to run road-cum-club competition machine) and been rewarded by a top reliability record; excellent handling within the limits imposed by standard suspension and no limited-slip diff; high comfort level with the "rallypack" bucket seats; good performance from the free-revving 1601cc engine and a fair amount of envy from friends who were getting rather tired of bigger, fatter, less agile saloons.

So I welcomed the Sport with open garage doors and a great feeling of well being. This was enhanced when I first sat in the car. The extra £105 spent on this model over the Mexico seemed well spent. Carpets, heated rear window, head restraints and twin door mirrors were all standard (and not found on the Mexico) though I was sad to note that such Mexico fittings as oil pressure and battery indicator gauges were missing. This change in equipment was my first clue to the different emphasis Ford were putting on the 1600 Sport as

opposed to the Mexico.

Because our car was a very early model, there was no chance to study an option list and tick the boxes. I thus had a rather expensive and not very logical list of extra-cost equipment.

In fact, my employers had been good enough to spend the sum of £115.72 on a black vinyl roof, push button radio (£63.47!), cloth trimmed seats and a red rear fog lamp, of which £14.91 for the seats would have kept me quite content. The lack of logic in all this became apparent when I found the seats had no recline facility as standard. Despite ordering cloth trim as an extra cost option, Ford want another £29.54 for the recline facility—which I find essential and many would expect to be inclusive

Facia is well designed (below) but has lost two gauges to the Mexico. Standard seats were changed to recliners (far right) which transformed the driving position and comfort level. Head restraints are standard on all Sport and Ghia models

in a car of this price. Left to my own devices, I would have been quite happy with spending just £45 on cloth, reclining seats. A lesson in option-watching if ever there was one!

Halfway through my spell with the car, however, we managed to swap the seats for recliners which confirmed my expectations : a few more degrees backward rake and everything falls into place—including my vertebrae.

First impressions

The Sport's exterior treatment is bright without being garish and although I feel the styling of all the new Escorts is too bland and anonymous around the rear, it's a body which is easy to live with.

I appreciated the easy access to the rear seats, the space available when I got there and the dramatically increased window area which is particularly valuable when reversing into a parking slot.

Two or three minor things were not up to par on delivery. A bright metal bracket inside one of the black front overriders (obviously a hangover from the days when all was chrome) had already rusted; a plastic collar round the base of the passenger head restraint was loose and would not go back inside the mounting without a struggle, and black streaks from the stick-on Sport stripes were discolouring the paint—particularly where the odd splash of petrol had rinsed the rear wing.

Within the first few hundred miles I realised the car's ride was very close to my taste, if a little bouncy for rear seat passengers over roads with any type of undulation. But there were already signs of running-on when I stopped the engine and a few rattles from the rear suspension.

Likes and dislikes

Apart from my obsession about the uncomfy angle of the standard seats there is little inside the Sport to criticise. If you wanted to carp then you might point to a limited amount of oddments stowage. Granted, there is a parcel shelf which is quite deep on the passenger side but for bits which should be easily accessible while driving (sunglasses, sweeties, cigs etc) there is nothing.

I would like to have seen a small centre console as standard with a tray for such things (the Triumph Herald had one!) and/or a small door pocket or flat, lipped area on top of the facia.

Most Sport owners will be pleased to note Ford's liberal use of grab handles over the windows —one for every passenger in fact AND the rear ones incorporate small hooks for clothes or hangers. A boon to a high mileage man who has to appear in a suit now and then! They may be less happy to note that the huge rear side windows do not open and this makes the rear seat more than a little sweaty in UK heatwaves despite the excellent airflow vents

up front. The latter have been changed for the good old Mk I Cortina eyeball variety which give superb volume and direction and must be very hard to beat at any price.

This brings me in a roundabout way to the minor controls which, like the curate's egg, are good in parts. The really minor ones, the tumbler switches for heated rear screen, hazard flashers and rear fog lamp are well placed and easily understood. But why should such a standardised item like Ford's simple heater control slides be exactly the opposite way round from the current Capri II? Illogical, as Mr Spock might say.

I would join him in the same description of the most-used steering column stalk, that for the indicators. On the new Escorts it's on the left of the wheel (where I prefer it), on the new Capris, it's on the right. The Escort is graced with two stalks on the right, the nearest and longest being for wipers (two speed) and washers, the farthest and shortest for side and headlights. No matter what the span of your hand, I defy you to reach this last one while still keeping a controlling grip on the wheel with your right thumb.

For normal road driving, particularly with a lot of city use where you are moving around inside the car during parking manoeuvres and leaping in and out a lot, I like inertia reel belts for convenience. The Sport's belts are quick and easy to use but seem grossly oversensitive to the angle of the car. If the car is parked on even a very slight uphill gradient, for instance, it's impossible to withdraw and fit the driver's belt.

Since the 1600 Sport has Mexico origins, comparisons are inevitable and though I like the new car a lot, I'm bound to say that it is less precise than our old stripey friend. Non-enthusiast drivers are hardly likely to notice these changes, but on my car, they are there. The brakes, even after servicing, have a long pedal travel and feel spongy. The "natural oversteer" has been ironed out and the car is very neutral, tending towards understeer. The steering has no feel, especially around the minimal lock position and there is little self-centring action. No coincidence, perhaps, that both Roger Clark and Billy Coleman found the same on their first outings with the basically similar RS 1800 and immediately ordered more castor angle.

The net result is that the car does not feel so eager and agile as the Mexico, though it sits more squarely on the road and is many degrees more civilised. Apart from obvious touches such as carpet in place of rubber matting, I appreciated refinements like a dipping rear view mirror, very good driving lights which come on with mainbeam, heated rear window, smart and useful door mirrors and a thick-rim, pleasant-to-handle steering wheel.

Finally, the maturity of this latest performance Escort is probably best illustrated by the

very low interior noise level—greatly reduced from Mexico days. Although the engine noise seems more subdued, it is wind noise and drumming from the road surface where the most dramatic improvements have been made. At 70 mph with the windows closed it is difficult to detect any wind noise and this has contributed a great deal to my enjoyment of the car over long distances.

Reliability

About halfway through my tenancy, the back axle started to whine at 45 mph or more. Gently at first, but getting noisier until 70 mph on the motorway became uncomfortable, the engine and most conversation constantly losing out to the ailing differential. I asked for it to be changed, but our dealer preferred to re-build it and solved the problem immediately.

The other main snag actually brought the car to a halt, but there's only me to blame as I should have foreseen it. I started to notice a slight gasp of in-going air as I released the fuel cap and realised the breather must be partly blocked. As I found later the little plastic pipe at the top of the tank had become totally kinked. My next 200 mile trip without a fuel stop ended with a strong smell of petrol in the boot. Sure enough, the constant pressure differential had been too much over such a long period of tank-emptying with no release of the cap, and the tank had partly collapsed !

In doing so, it had slipped a connection to the fuel line and around half a gallon of petrol had leaked out under the tank into the bottom of the offside rear wing. Thankfully, the tank hadn't ruptured and while I raced off somewhere to report a rally, a friendly garage near my home dismantled everything, popped the tank back into shape with a compressor and air line and re-fitted tank, line and seal—all for £1.33.

The very low-slung exhaust might have caused problems from the outset on some of the undulating minor roads I keep finding, but at my request the dealership re-slung it and won about 3in of ground clearance with very little trouble.

Ford recommended mods to our car comprised a change of front brake hoses, an obscure mod on the door locks and a change of plugs from Autolite AG 22 to AG 12 (not the racing AG 12!). I hoped the latter would cure a chronic running-on problem which developed with the car and at first it seemed to, but this now returns midway between each service and I'm told the distributor seems to advance itself regularly through rapid points wear. Things don't seem to get too hot and the Sport is quite happy on four star fuel. It is a willing starter on the new automatic choke.

The car had done very few miles when first one side, then

the other rear suspension started rattling and producing weird banging noises even on gentle take-offs. The dealer admitted to tightening the upper rear shocker mountings and nothing more: all was cured.

Lastly, the exhaust manifold (with all its tortuous and non too neat welding) was changed under guarantee after starting to blow and rob the car of much acceleration.

What went wrong ?

Mlge	Fault
22	Loose plastic collar on passenger head restraint
	Rust on bright metal bracket inside one overrider
	Gearchange stiff on 2nd to 3rd movement
500	Engine running-on badly when warm
	Rear suspension rattles on take-off
	Boot lock protrudes when key is withdrawn
3500	Running-on now pronounced (Ford recommended plug change seemed to cure for a few hundred miles)
	Rattles from rear became worse and dampers had to be retightened
4350	Fuel tank breather became blocked and caused tank to collapse partly during long journey, resulting in petrol leaking into boot
	Very slight water leak from windscreen, starts and stops of its own accord
	Brake hoses changed on Ford's recommendation
	Whining differential cured by re-build at dealers
8114	Running-on problem returns
	Accident involving stray dog entails £67 repair bill, £12.83 for a replacement grille
10688	Exhaust manifold changed in successful bid to trace slight leak and consequent lack of performance

Running costs

One thing that strikes us about the Escort's economics is its willingness to turn in really good consumption figures when in a good state of tune.

After an understandable loosening-up period, it consistently achieved over 30 mpg in a cross section of London traffic and motorway cruising work until the tune started to slip. As can be seen, this knocked the figures down to around 25 mpg at one point, though a blowing exhaust manifold aggravated the situation caused by our more usual culprit —the distributor.

Tyre and oil costs are both reasonable and we class the servicing costs as very reasonable. The only bad news was the high cost of bodywork repairs after a contretemps with an Irish farm dog (which lived to dash out on someone else). What appeared at first to be a slightly dented front panel and number plate turned into a bill for £67.46 after stripping, straightening, re-painting, re-stripping, re-fitting et el had been done. Mostly, we begrudged £12.83 for a new plastic grille because of a 2in split in the original. Had this been good old aluminium, we would have hoped a slight crease could have been gently ironed out.

Conclusion

The Sport was quieter, more refined, smoother and a good deal more socially acceptable than its ancestors. But there is no getting away from the fact that it is less of an enthusiasts' car than the Mexico. It doesn't impart the same feel, is not so taut and cannot be driven quickly through winding lanes with the same precision.

While greatly enamoured of its reliability the Sport's failure to stay in tune and a constant lack of "edge" to its performance has annoyed me. The bouncy ride experienced by rear seat passengers also goes against it as a family car.

I quite like the Sport but, to be honest, it rather falls between two stools in my view: it's neither a tummy-tingling sporty saloon nor a frill-free family holdall. I'd be happier with better handling, a little more power and a more logical division of standard and optional equipment.

Second opinion

I had misgivings about swapping my Lancia Beta for David's Escort Sport for the long 800-mile round-trip to Ingliston. Second opinion drives don't have to be *that* long. I knew that the Lancia, with its good ride, superb seats and easy cruising would get me to Scotland without fatigue, aches or pains. Perhaps with the old Mexico in mind—a good fun car but hardly a long-distance express —I had my doubts about the Sport.

On the whole, they proved misguided even though neither the seats nor the ride are quite as comfortable as the Lancia's. But for what's now virtually a non-stop motorway run, the Sport was comfortable enough, excellently ventilated (the ability to deflect cool air on your face from the screen vents with the sun visors is a refreshing bonus that Ford underplay), and certainly much lighter work than the Lancia, particularly in the steering and gearchange departments. It was nice to relax and steer almost by telepathy rather than muscle power. The car ran sweetly and fairly quietly (especially on the way back when there was no wind to buffet the car and increase roar) at about 4000 rpm most of the way, averaging over 31 mpg— which is better than the Lancia would have done at the same average speed. Moreover, the Sport needed only one mid-way stop to refuel: the Lancia would have needed two.

Apart from a flat spot at very low revs and various minor rattles, the car was in excellent shape. I rather liked it.

Roger Bell

12,000 MILE TEST • Ford Escort 1600 Sport

Make : Ford
Model : Escort 1600 Sport
Makers : Ford Motor Company, Warley, Brentwood, Essex
Price : £1860

WHAT IT COST

Petrol (1)	£306.02
Servicing (2)	£45.77
Oil (3)	£1.80
Tyres	£17.00
Road fund licence	£40.00
Total	£410.59

(1) 29 mpg at 74p per gallon.
(2) Three services*—the first of which was "free" but actually included £4.62 labour for cleaning rear brake adjusters and bleeding the system after we complained of a spongy pedal.
(3) Excluding oil changes. Consumption averages out to around 1,000 miles per pint.
* Parts fitted : three sets points; three sets plugs ; one rocker gasket ; two oil filters; two air filters; one exhaust manifold.

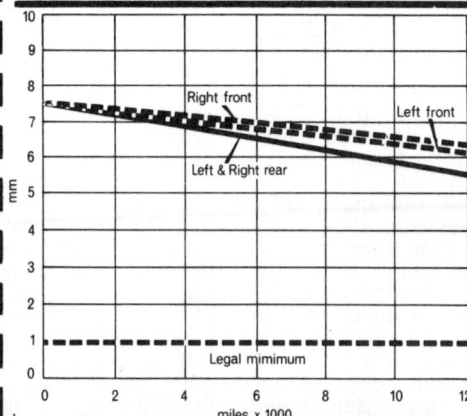

The Michelin ZX tyres fitted to the Sport (165 x 13) are set for a long and healthy life — in fact 36,000 miles for the faster wearing rears. The car suffered one easily repaired puncture from a piece of scrap metal on the M1

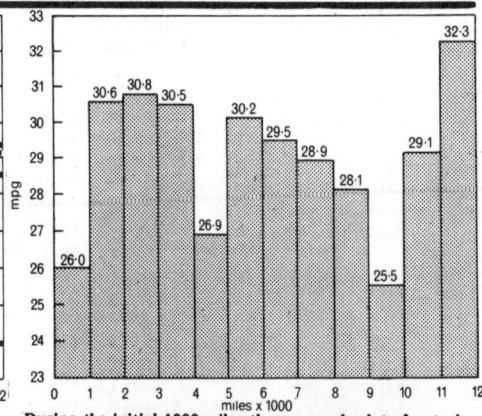

During the initial 1000 miles the car used a lot of petrol, although we restricted ourselves to 4000 rpm max. Things then stabilised until the points wore out at 4000 miles. The gradual decline from 6000 represents the car again going out of tune, culminating in a low when the exhaust manifold was also leaking. The final figure shows what can be done on a 70 mph cruising journey, London-Scotland and return

PERFORMANCE

CONDITIONS
Weather	Fine, wind 0-5 mph
Temperature	76-80°F
Barometer	29.6 in Hg
Surface	Dry

MAXIMUM SPEEDS
	Our car mph	Test car mph
Banked circuit	97.9	97.9
Best ¼ mile	100.0	100.0

Terminal Speeds :
at ¼ mile	74
at kilometer	87

Speed in gears (at 6,500 rpm) :
1st	36	36
2nd	61	61
3rd	85	85

ACCELERATION FROM REST
mph	sec	sec
0-30	3.6	3.3
0-40	6.0	5.3
0-50	8.4	7.9
0-60	12.0	11.1
0-70	15.4	15.5
0-80	23.6	22.9
Standing ¼ mile	18.7	18.2

ACCELERATION IN TOP
mph	sec	sec
20-40	12.0	9.9
30-50	11.0	10.6
40-60	10.1	10.0
50-70	10.8	11.0
60-80	13.0	14.4
70-90	18.5	—

FUEL CONSUMPTION
Touring*	34.4 mpg
	8.2 litres/100km
Overall	29 mpg
	9.7 litres/100 km
Fuel grade	97 octane
	4 star rating
Tank capacity	9 galls
	40.9 litres
Max range	309 miles
	497 km
Test distance	12,010 miles
	19,324 km

* Consumption midway between 30 mph and maximum less 5 per cent for acceleration.

SPEEDOMETER (mph)
Speedo	30 40 50 60 70 80
True mph	31.5 41 51 60 70 79
Distance recorder : 2 per cent fast.	

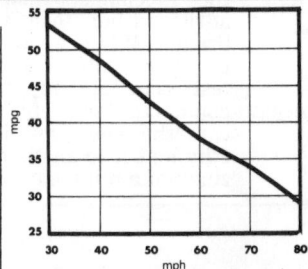

The Escort's good economy potential shows up well here. 30 plus mpg at 79 mph is reflected by our final figures in the 12,000-mile test, in which a long steady journey played the main part

WEIGHT
	cwt	kg
Unladen weight*	17.0	863.6
Weight as tested	20.7	1051.6

* with fuel for approx 50 miles.
Performance tests carried out by Motor's staff at the Motor Industry Research Association proving ground, Lindley.

COMPARISONS

	Capacity cc	Price £	Max mph	0-60 sec	30-50* sec	Overall mpg	Touring mpg	Length ft in	Width ft in	Weight cwt	Boot cu ft
Ford Escort 1600 Sport‡	1598	1860	97.9	12.0	11.0	29.0	34.4	13 0.5	5 0.5	17.0	8.9
Alfasud Ti	1168	1868	99.1	12.9	15.6	26.4	32.5	12 10	5 3	16.5	9.4
Colt Lancer SL 2-door	1597	1849	95.8	11.6	10.7	26.8	35.7	13 1	5 0	16.3	7.6
Datsun 180B SSS	1770	2286	104.7	10.3	9.0	24.7	31.4	13 10.75	5 2.75	20.1	9.6
Fiat 124 ST	1592	2493	99.8	10.1	13.1	25.4	29.8	13 2.75	5 4.5	18.3	10.0†
Morris Marina 1.8 TC	1798	2064	99.3	11.8	9.3	24.7	28.8	13 7	5 4	18.4	12.4
Opel Ascona SR	1897	2360	98.7	10.5	9.3	25.2	28.9	13 8	5 5	18.7	10.5
Austin Allegro HL	1748	2095	99.7	10.0	9.3	25.6	31.6	12 8.25	5 3	17.0	9.2

*in top. Fifth for Colt, Alfa, Fiat and Austin. Fourth for others   †measured with boxes, not cases
‡12,000-mile car

RS1800

HOW POPULAR CAN YOU GET?

Many people thought that the closure of AVO would spoil Ford's limited production RS models, but the RS1800 and RS2000 (with Firenza-style droop snoot) have disproved that theory. We tried the 1800 recently (the bigger-engined car has still not been released) and found it just as well engineered as its predecessors, even if it lacked some of the tautness in handling which was a feature of the old RS1600.

Actually, you're very lucky to be able to read a test of the RS1800 here. The first one we tried dislodged its gear linkage in the wilds of Surrey one Saturday night, just a week after Motor's road test of the same car appeared. The writer said 'No matter how brutal we were with our upward changes . . .', but to be fair the mechanism had come adrift rather than actually broken. While we waited to get back into Ford's road test schedule so that we could take some pix of the car, the company ill-advisedly lent it to another magazine not a million miles from

Lukewarm Car and it finished up smashed to pieces at the MVEE offroad course down in Chobham.

Finally we got a different car, a sober red one which lacked the go-faster stripes on the wrecked version, and which seemed rather better in several respects. It was slightly quicker, didn't make any funny induction noises, and didn't suffer from a slight misfire.

One of the main differences between the RS1600 and the 1800 apart from the new bodyshell is in the price: admittedly there's been a lot of inflation lately, but all the new Escorts, with the exception of the Popular, are somewhat ugly in the price department. This little racer, for instance, will break your bank manager's heart (yes, even they have feelings) to the power of three thousand if you take the full Custom Pack—and is that a breach of copyright or what?

The body of the RS1800 is standard Halewood tin except for the spoilers; the front one is pop-riveted, while the boot device is made of squishy rubber so

that you won't crush any hedgehogs if you're brainless enough to roll the car on an offroad course.

The engine is basically a bored out RS1600 unit, which was in turn an adaptation of the Cosworth F2 engine. Capacity is now 1845cc, and while the twin overhead cams and the valve gear are identical with those of the earlier car, the use of a single 32/36 Weber in place of the pair of side-draught ones has actually reduced the horsepower rating by a small amount. But the torque figure has increased, and though there's a small loss in terms of outright acceleration and maximum speed, the increased flexibility makes it a more sensible road car.

At just under 9.0sec, the 0-60mph time is about half a second slower than that of the RS1600 and the top speed of about 111mph is only two or three mph down, so the single carb doesn't carry a very heavy penalty. Fuel consumption has improved dramatically, and without soft-pedalling you can still get over 30mpg fairly easily, which is excel-

lent in view of the car's power.

The gearbox is of German origin, and was designed for Pintos and smaller Mustangs. But it's every bit as positive as the boxes used on earlier RS models (except when it leaves you without any gears in the middle of Surrey on a wet Saturday night), and the clutch is well matched to it.

The RS1600 and the Mexico gave Ford a justifiable reputation for making performance cars with superb roadholding and handling. For some reason, in recent years their specialist cars have tended more and more towards excessive understeer; this was noticeable in the old RS2000 (and presumably the new one) and also in this RS1800. It's probable that they've done it intentionally as a 'safety factor', but it's something that I for one could well do without, and I imagine a number of owners will set to work modifying the front anti-roll bar and so on to get back to neutral ground. This kind of messing about shouldn't be necessary for three grand.

Presumably it's a tribute to the designer of the spoilers to say that this car is much more stable in a straight line than its predecessors, for passing big trucks on motorways on windy days causes little or no deflection from one's desired course. The steering too, is highly wonderful, giving lots of feel and lacking in heaviness even at low speeds.

The brakes—the familiar discs/drums with dual circuits—are just as excellent as ever, progressive in action, immune to fade, and never pulling to one side. Pitch under heavy braking is also kept to a minimum, unlike in a fair number of today's sporty saloons.

While I didn't go bananas over the handling, for reasons stated above, I can certainly make no complaint about the roadholding. The 175/70 Pirelli CN36 tyres on 5½J steel sports wheels cling on well in both wet and dry. The combination of stiff dampers and soft springs makes changes of attitude in the car nice and progressive, and also provides an exceptionally good ride,

which gets better the quicker one goes.

The interior is, as always, well trimmed, with a good driving position, a good selection of legible instruments and, as in all the new Escorts, very good all-round vision. I found the seats a bit uncomfortable though, and while the rear bench isn't of the legless midget variety, it isn't designed for giants either. The cloth trim on the seats and the deep pile carpet are nice touches of luxury.

Interior noise is quite reasonable. Even at high revs the engine doesn't become too obtrusive, and all other noises are kept to a reasonable level, though there's a fair amount of wind noise at high speeds.

The main thing that the RS1800 proves (apart from the fact that I can't afford one) is that Ford can still turn out good specialist cars off their ordinary production lines. The cars are taken off the main line at Halewood and finished off on a subsidiary one, and the results are very good indeed. But not to the power of three thousand. **PD**

Ford Escort RS2000

Plastic-fronted, Cortina-engined Escort with a little aerodynamic help offers excellent performance and value for money.
Taut character to the handling and ride.
Brakes rear-biased on test car.
Superbly comfortable and locating rally-type seats in front at expense of rear occupants.
Range limited by standard Escort tank. Great fun.

FORD ARE not exactly strangers to making their top-selling Escort go faster. Both the original Escort and its current form have, in more powerful versions, proved international rally winners, as Timo Makinen's magnificent victory in November's Lombard RAC Rally once again demonstrated.

The Makinen car was, of course, a highly-developed 240 bhp group 2 and 2-litre version of the BDA 16-valve-engined RS 1800, which continues as the car for rallying enthusiasts. The subject of this test in no way replaces or supplants the RS 1800, and although its special front-end treatment, general appearance and performance may suggest a similar purpose, it is in fact not a competition-intended Escort.

It is built by Ford Saarlouis and first appeared at Earls Court in 1975. The engine is the belt-driven single-ohc, 1,993 c.c. unit used in the Cortina originally, but uprated from 98 bhp at 5,500 rpm to 110 bhp (DIN) at the same crankshaft speed. This is achieved by a new cast exhaust manifold and different exhaust system; torque is

AUTOCAR, w/e 17 January 1976

raised, from 111 lb. ft. at 3,500 rpm to 119 lb. ft. at 4,000 rpm. The gearbox is Cortina, with a shorter gearlever, and therefore more precise gearchange, joined to the engine by a different cast light-alloy bellhousing, which, with a similarly made sump is claimed to reduce noise and vibration levels. The final drive is a 3.54-to-1, compared with the Cortina's 3.75. On Ford/GKN cast light-alloy 6in. rimmed wheels, which are standard, and Pirelli Cinturato CN36 steel-braced radial tyres of 175/75HR-13in. section the car is geared at 18.6 mph per 1,000 rpm.

Suspension settings are slightly different to try and provide what Ford believe to be the right compromise between ride and handling. At the back, the normal anti-roll bar has been replaced by twin trailing arms, as extra location for the leaf-sprung live axle.

A slight attempt has been made to improve aerodynamic drag. The 6in. extended nose has a small air dam, and is slightly raked if you ignore the effect of the grille. It is entirely moulded in polyurethane, which is flexible and therefore rather safer from a pedestrian's point of view, or contact. Ford claim a "significant improvement" in streamlining and front-end lift, plus an "even greater" reduction in rear lift by virtue of the moulded rubber boot-lid spoiler. A remarkable feature for a production car is the design of the front seats, which are more the sort of thing you find in a rally car — high-backed, adjustable-raked, and with marked sideways support.

There will be buyers who may assume, and salesmen, possibly who may assure, that the RS 2000 is a rorty rally car. It is easy to assume that from the appearance and the comprehensive equipment. In fact, we have a suspicion that Ford would do well to offer the RS 2000 in more plain paint schemes as well as the present heavily-flashed one, since some of the "kind of sporting buyers wanting performance with refinement in a compact, yet practical saloon" which they say the car is aimed at may well prefer something less conspicuous.

Performance

The performance cannot help being somewhat conspicuous, by virtue of its magnitude rather than its noise, for in acceleration at least this is about the fastest in its class, as our comparison selection shows. Ford claim 111 mph as top speed, which is bettered by other cars; the 109 mph mean returned on a somewhat windy MIRA day is equivalent to 5,850 rpm, which as it is 350 rpm over the engine's peak power speed, suggests that slightly higher gearing would be possible. We certainly thought that an appreciably higher, overdriving fifth would be very worthwhile, not to increase top speed — if it were a true overdrive, it wouldn't anyway — but to lessen the slight fussiness of the car at motorway speeds. It cruises at an easy 90 mph, and would do so appreciably more easily with higher gearing. However, at such speeds there was not too much wind noise on the test car, and listening to the radio had

The RS 2000, in the example tested, could be provoked into rear-end slides on MIRA's test track with some ease. Note how even here roll is kept well down; at public-road cornering speeds, this is one reason why the car has such a reassuringly taut feel

Ford Escort RS2000

not become difficult. Whilst on the subject of noise, it must be said that the car suffered from some disappointing body boom at low speeds in the otherwise perfectly flexible top gear, both on drive and overrun, at around 25 mph. There is also a tiresome amount of boom when passing over coarse surfaces; Ford tell us that they hope to improve this on cars sold to the public. We noted too a small tendency to heterodyne at 60 to 70 mph.

In all other respects the engine and transmission are delightful. The power unit is generally smooth, with no irritating flat-spots, demonstrating that *some* manufacturers can arrange car-buration to suit a sports engine which also complies with coming European exhaust emission restrictions. A touch of the throttle sends the revs spinning up; com-bined with the quickness of the terse, narrow-gated, short-travel gearchange, one can play the slickest of tunes up and down the box. The excellent pedal layout, perfect for heel-and-toe, helps here too. Some of us found the change a little notchy, though not all. Ford describe the Cortina ratios as "close"; the gaps between the top three gears are not wide, but first is a little low.

For that reason, during the acceleration runs we found it best to change up from 1st to 2nd at rather more than the 5,800 rpm change point which we set our-selves for the subsequent changes. The point here is that although there is neither a rev-limiter or a red line on the revcounter, so that it is possible even in 3rd gear to rev to the 7,000 rpm end of the scale, no useful purpose is served by going higher than 5,800 rpm, since there is an obvious fall-off in acceleration. Ford tell us that 6,600 rpm is in fact

the maximum recommended, which corresponds to the speeds given in the table; 5,800 is equiva-lent to 30, 55 and 79 mph in the intermediates.

Best standing starts were made by dropping the clutch in at 4,000 rpm, which produced just the right amount of wheelspin for some excellent getaways. The figures speak for themselves; 50 in 6.4 sec, 60 in 8.6, the standing quarter-mile in 16.7, and 100 in 33.6. The RS2000 accelerates most satisfactorily, and will comfortably see off the major-ity of its opposition. Of the obvious competitors, several are faster in top speed, but only the Dolomite Sprint, which weighs 150 lb more than the Ford's trim 18½cwt is able to offer any comparable go.

It starts easily, though the auto-matic choke has to be kicked out of action with a dab on the throttle before it cuts itself out on its own. One is conscious of a pleasant level of engine noise, with a hint of a crackle in the note between 2,200 and 2,500 rpm.

Economy is not a strong point. The best interval consumption we saw was 25 mpg, which is almost exactly what the DIN figure based on the steady speed graph works out at. The car's thirst is made irritatingly obvious by the silly little nine gallon tank, which gives a maximum range of only a little more than 200 miles. The car is certainly worth something more generous than the standard Escort tank; 12 gallons would start to be more appropriate. As usual with this engine, oil consumption was very low indeed — a Ford strong point.

Handling, ride and brakes

One's main reaction to the RS 2000 is to describe it as a pretty taut feeling car. It doesn't roll much, and response to movements of the

small-ish 14in. dia three-spoke steering wheel are almost as imme-diate as on a competition saloon. Some people will perhaps complain that the steering is a little heavy, at manoeuvring speeds particularly, but we would not want it lightened. It self-centres much better than one has come to expect from Fords, gives good feel of what the front tyres are doing, and has a quite tight minimum turning circle (31ft. 3in. mean). Catching the almost inevitable tail slides provoked on wet roads by a clumsy or over-eager throttle foot is pleasingly quick.

Of the overall handling of the test car when it is hurried, we were not at first so completely sure — some of us at any rate. Although it does not suffer from the rear anti-roll bar which causes too-easy inside-rear-wheel lifting on other Escorts, there is a deliberately aimed-for tendency towards neutral steer as the car starts to roll during harder cornering which can at first alarm some drivers. Cornering progressi-vely harder through a long bend will make one rear wheel spin with power; throw the car at the bend — on a closed track of course, and it starts to go neutral; lift-off abruptly as well and the tail will slide nicely. On the MIRA banking at top speed it accordingly felt a bit twitchy, as you might expect — roll effects were making themselves felt — although in fact the car was per-fectly stable, more so than it seemed.

This does not alter the fact that, on public roads, the RS 2000 is a very quick cross-country car, once you have got used to its slightly unusual feel. It moves somewhat too noticeably in a side wind. The ride is sporting, which is a polite way of saying it is bumpy — bumpy but very well damped in a way that makes it quite acceptable.

Stalks are laid out Continental style— signalling on left, lamps and two-speed-plus-hesitation-wipers and wash on right. Nacelle carries exemplarily clear rev counter (left) and speedo with mph and kph markings (right), with fuel, oil pressure and water temperature in middle above strip of warning lamps (hidden). Note useful BMW-style centre cubby box and clock to right of open dash-cubby

Specification

ENGINE

Cylinders	4, in line
Main bearings	5
Cooling	Water
Fan	Fixed
Bore, mm (in.)	90.82 (3.57)
Stroke, mm (in.)	76.95 (3.02)
Capacity, cc (in³)	1,993
Valve gear	Ohc
Camshaft drive	Toothed belt
Compression ratio	9.2-to-1
Octane rating	97 RM
Carburettor	Twin-choke Weber
Max power	110 bhp (DIN) at 5,500 rpm
Max torque	119 lb ft at 4,000 rpm

TRANSMISSION

Clutch		
Gear	Ratio	mph/1000 rpm
Top	1.0	18.6
3rd	1.37	13.6
2nd	1.97	9.4
1st	3.65	5.1
Final drive gear	Hypoid bevel	
Ratio	3.54-to-1	

SUSPENSION

Front —location	MacPherson strut
springs	coil
dampers	telescopic
anti-roll bar	Yes
Rear — location	Radius arms and leaf springs
springs	half elliptic
dampers	telescopic
anti-roll bar	No

STEERING

Type	Rack and pinion
Power assistance	No
Wheel diameter	14 in.

BRAKES

Front	9.6 in. dia disc
Rear	9in. dia x 1.75 in drum
Servo	Yes

WHEELS

Type	Cast light alloy
Rim width	6 in.
Tyres — make	Pirelli
— type	CN36
— size	175/70 HR-13in.

EQUIPMENT

Battery	12 volt 57 Ah
Alternator	45 amp, 220/120 watt
Headlamps	Cibié Halogen 4 lamp system
Reversing lamp	Standard
Hazard warning	Standard
Electric fuses	7
Screen wipers	2-speed
Screen washer	Electric
Interior heater	Air blending
Interior trim	Cloth seats, pvc headlining
Floor covering	Carpet
Jack	Screw pillar
Jacking points	One each side under sills
Windscreen	Laminated
Underbody protection	Zinc phosphate electrocoat and pvc under wheelarches

MAINTENANCE

Fuel tank	9 Imp galls (41 litres)
Cooling system	12 pints (inc heater)
Engine sump	6.6 pints SEA10W/50
Gearbox	2.6 pints SAE80EP
Final drive	2 pints SAE90 Hypoid
Grease	No points
Valve clearance	Inlet 0.008 in (cold) Exhaust 0.010 in. (cold)
Contact breaker	0.025 in. gap
Ignition timing	8 deg BTDC (static)
Spark plug — type	Motorcraft BF32
gap	
Tyre pressures	F24; R 22 psi (normal driving)
Max payload	801 lb (445 kg)

AUTOCAR, w/e 17 January 1976

Maximum Speeds

Gear kph	mph		rpm
Top (mean)	109	175	5,860
(best)	112	180	6,020
3rd	90	145	6,600
2nd	62	100	6,600
1st	34	55	6,600

Acceleration

True mph	Time secs	Speedo mph
30	3.0	30
40	4.7	40
50	6.4	50
60	8.6	60
70	12.7	71
80	16.9	82
90	23.1	93
100	33.6	104
110	—	116

Standing ¼-mile:
16.7 sec — 80 mph
Kilometre:
31.7 sec — 98 mph

mph	Top	3rd	2nd
10-30	—	6.6	3.9
20-40	8.7	5.8	3.6
30-50	7.8	5.3	3.5
40-60	7.8	5.1	4.5
50-70	8.4	5.9	—
60-80	9.1	7.6	—
70.90	10.6	13.2	—
80-100	16.9	—	—

Consumption

Fuel
Overall mpg 24.7
(11.4 litres / 100km)
Calculated (DIN) 25.1
(11.2 litres / 100km)

Constant speed:

mph	mpg
30	38.1
40	38.8
50	35.4
60	32.3
70	27.6
80	24.8
90	21.2
100	16.7

Autocar formula:
Hard driving, difficult conditions 22.5 mpg
Average driving, average conditions 27.2 mpg
Gentle driving, easy conditions 32.1 mpg

Grade of fuel: Premium, 4-star (97 RM)

Mileage recorder:
3.8 per cent over reading

Oil
Consumption (SAE 10W/40) negligible

Brakes

Fade (from 70 mph in neutral)
Pedal load for 0.5g stops in lb

	Start/end		start/end
1	42/50	6	45/50
2	40/45	7	45/50
3	37/42	8	45/50
4	37/42	9	48/50
5	45/50	10	42/50

Response (from 30 mph in neutral)

Load	g	Distance
20lb	0.16	188ft
40lb	0.47	64ft
60lb	0.75	40ft
80lb	0.90	33ft
100lb	0.92	32.7ft
Handbrake	0.32	94ft

Max. gradient 1 in 3

Clutch Pedal 28lb and 5¼in.

Test Conditions

Wind: 10-20mph
Temperature: 9 deg C (48 deg F)
Barometer: 29.8 in. Hg
Humidity: 65 per cent
Surface: dry asphalt and concrete
Test distance: 906 miles

Figures taken at 2,300 miles by our own staff at the Motor Industry Research Association proving ground at Nuneaton

All Autocar test results are subject to world copyright and may not be reproduced in whole or part without the Editor's written permission

Regular Service

Interval (miles)

Change	6,000	18,000	36,000
Engine oil	£1.98	£1.98	£1.98
Oil filter	£2.69	£2.69	£2.69
Gearbox oil	—	—	—
Spark plugs	£2.60	£2.60	£2.60
Air cleaner	—	£2.75	£2.75
C/breaker	£0.58	£0.58	£0.58
Total cost	**£20.75**	**£26.51**	**£31.24**

(Assuming labour at £4.30/hour)

Parts Cost

(Including VAT)

Brake pads (2 wheels) — front £8.97
Brake shoes (2 wheels — rear £4.92
Silencer (s) .... Price not yet available
Tyre each (typical advertised £21.98
Windscreen £24.32
Headlamp unit . Price not yet available
Front wing £14.69
Rear bumper £11.55

Warranty Period
One year, unlimited mileage

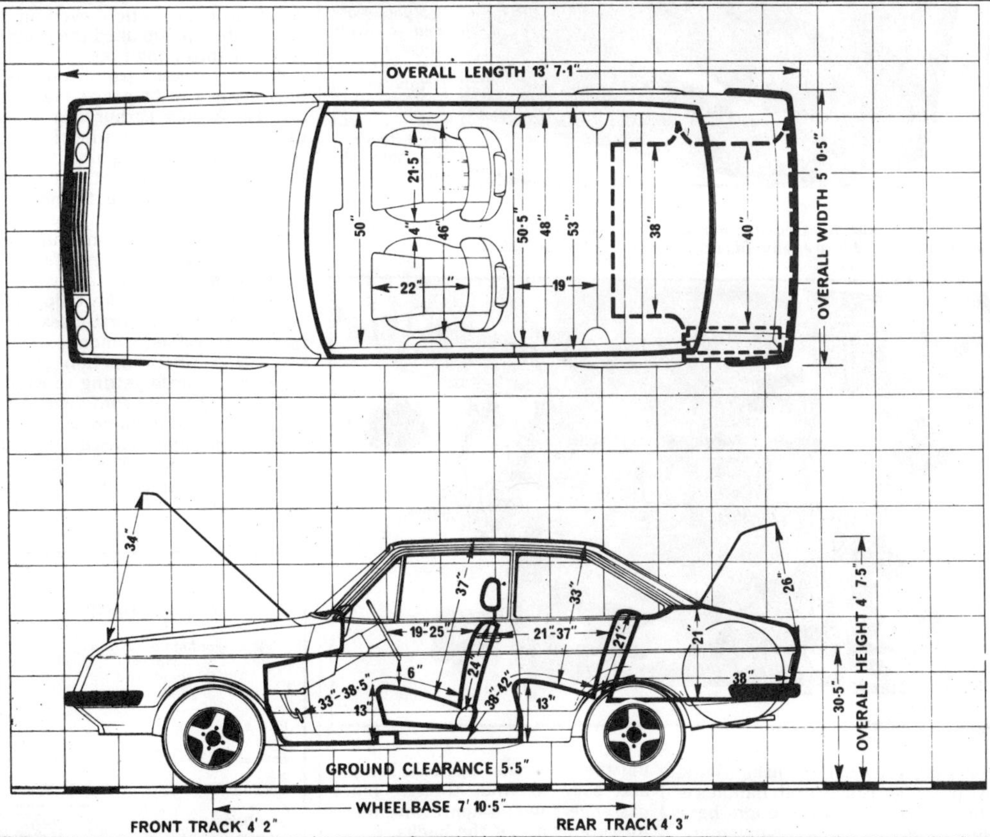

Weight

WEIGHT:
Kerb, 18.5cwt / 2,075lb / 941kg
(Distribution F/R, 56.0/44.0)
As tested,
22.2cwt / 2,485lb / 1,127kg

Boot Capacity: 15.8 cu.ft.

Turning Circles:
Between kerbs
L, 31ft 4in; R, 31ft 2in
Between walls
L, 32ft 8in; R, 32ft 6in
Turns, lock to lock 3.5

Test Scorecard

(Average of scoring by *Autocar* Road Test team)

Ratings: 6 Excellent
5 Good
4 Better than average
3 Worse than average
2 Poor
1 Bad

PERFORMANCE 4.00
STEERING AND HANDLING.... 3.42
BRAKES 3.20
COMFORT IN FRONT 4.08
COMFORT IN BACK 2.57
DRIVERS AIDS 4.13
(instruments, lights, wipers, visibility, etc)
CONTROLS 3.88
NOISE 4.00
STOWAGE 4.00
ROUTINE SERVICE 3.10
(under-bonnet access: dipstick, etc)
EASE OF DRIVING 3.64
OVERALL RATING 3.62

Comparisons

	Price £	max mph	0-60 sec	overall mpg	capacity c.c.	power bhp	wheelbase in.	length in.	width in.	kerb weight	fuel gall	tyre size
Ford Escort RS2000	**2,857**	**109**	**8.6**	**24.7**	**1,993**	**110**	**95**	**163**	**63**	**2,075**	**9.0**	**175/70-13**
Alfretta GT	4,198	117	9.4	23.7	1,779	122	95	165	65	2,315	11.8	185/70-14
Audi 80GT	2,910	106	9.5	29.5	1,588	100	97	165	63	1,881	10.0	175/70-13
BMW 320	3,349	108	10.2	21.2	1,990	109	101½	171½	63½	2,271	11.5	165-13
Toyota Celica GT	—	113	9.3	29.8	1,588	124*	95½	164	63	1,955	11.0	165.13
Triumph Dolomite Sprint	3,083	115	8.7	23.6	1,998	127	96½	162	62½	2,228	12.5	175/70-13
Vauxhall Magnum 2300	2,224	103	10.0	25.3	2,279	110	97	163	64¾	2,155	12.0	175/13

*SAE

Ford Escort RS2000

The brakes on the test car were disappointing, exhibiting all the signs of badly bedded-in friction materials, causing early rear locking. Fade resistance is good however, and the handbrake coped well with our 1-in-3 test slope; the car took off easily, with lots of wheelspin when restarted here.

Driving position and controls

You sit somewhat high in the RS 2000, compared to a normal Escort; one has to be careful to duck slightly more than normally when getting in or out if tall, to avoid clouting one's head on the door frame. Both driver's and front passenger's seat are much more like proper rally seats than usual, which is an excellent thing in a car of this character. They are big seats too, big in overall size, so that we suspect that their maximum rearward movement has been restricted to allow some room behind. This does not matter however in front, because, most wisely, the seat designers have shaped the squab so that it fits a rather knees-up posture very comfortably indeed, with just the right support for the thighs — for tall men; shorter legged drivers found reaching the pedals awkward.

British drivers will rightly object to the otherwise excellent minor control layout, since it has the signalling/dip stalk on the Continental side, with the wipers and lamp swtiches opposite. Our only other plea is for horns to be sounded from the middle of the steering wheel instead of the indicator stalk; standardisation of controls is coming, and since a central horn push remains easily the quickest to find in an emergency, this is one now-Continental arrangement we would like to see made universal.

Instruments are very well laid out and marked, living in what one now recognises as a good German arrangement, a pod in front of the driver, shielded from the windscreen, and with a glass that does not pick up unwanted reflections. The trip mileometer has the simplest of push-zero buttons. There is an oil pressure indicator, and, on the test car, a fuel gauge that would not read "full" when it should. For the first time, the two-speed wipers on this Ford have an additional delay wipe setting, with a 7sec pause between each stroke — invaluable in this country.

In front of the passenger there is a clock, angled slightly towards the driver, and a cubby hole and glove box. It is good to see on this model at any rate that Ford do acknowledge the existence of drivers who want somewhere to put easily-reached oddments, like a camera, maps, or pencils and so on. The centre console in front of the gearchange includes some very useful extra storage space, as well as a place for any radio fitted.

The front passenger has a vanity mirror in the visor. Neither visor can be swung sideways to shield

Above: The RS 2000's unique paint scheme and shape make it stand out from other Escorts

Left: The British-designed and made rally-style seats hold a wide range of drivers as comfortably and as well as they look — a great asset to driving both far and fast

Unusual blue-plastic air cleaner and silencer is squashed over cam cover and Weber carburettor. The 2-litre engine fits fairly easily by shoehorn standards

one from a low sun on either flank. All passengers have grab handles in the roof sides, and the doors have small armrests that are convenient.

The halogen headlamps are superb on main beam, giving a tremendous range with useful spread. Visibility on current Escorts is usefully better than on the previous model, with no serious blind spots on any quarter.

Getting into the back on this two-door model is not much worse than on other similar cars. Releas-

ing the front seat backs is simple and easy to find, though older passengers would appreciate more forward swing of the backs. If the driver is average tall, and is selfish about his own comfort, rear occupants suffer accordingly, with limited knee space, although there is acceptable foot room. They will also find their heads may need to be stooped a little, since they are forced to sit well back and upright. The considerate driver taking three people out in the RS 2000 has

therefore to sacrifice a notable amount of his own space.

In the back there are small side ashtrays and a handy, vee-bottomed window shelf. The quite deep windows prevent one otherwise feeling shut in unduly.

Heater behaviour is not quite as good as we used to expect from this manufacturer. As on other current Escorts, there is far too little of the temperature control lever's movement that actually varies the temperature. Only very near the hot end of the scale can one vary the temperature of the screen vents, and only correspondingly near the cold end can one vary that of the footwell ones. Conversely therefore, one tends to roast one's feet whilst general head-level heat is almost too cool. There is good through-flow however, and the usual unheated air side vents which worked adequately, in the cold weather of the test period.

Because they cannot help adding unwanted weight and complication, we do not care for American 5 mph bumpers. But the flimsy piece of trim attached by similarly flimsy light gauge brackets to the tail which is presumably meant to pass for a bumper on this car goes too far in the opposite direction. Owners will do well to bear in mind that this ephemeral protrusion will be of little protection even in a parking bump. We liked the practical black finish however.

Engine access isn't too bad. Most items can be got at reasonably. The Weber carburettor hides under an unusual plastic air-cleaner and trunking unit. which presumably because of the small bonnet space has to be somewhat squashed over the engine. The dipstick is obvious to find, and the distributor has an anti-interference shield round it.

Overall, this is a highly enjoyable Ford, representing first-class performance value-for-money. The test car, a pre-production prototype, took a little getting to know before one became confident in it. We expect production models to be more immediately reassuring, and just as much fun. This is a very satisfying motor car of encouraging potential.

MANUFACTURER:

Ford Motor Co. Ltd.,
Warley
Brentwood
Essex

PRICES	
Basic	£2,441.88
Special Car Tax	£203.49
VAT	£211.63
Total (in GB)	**£2,857.00**
Seat Belts	£27.27
Licence	£40.00
Delivery charge (London)	£35.00
Number plates	£6.48
Total on the Road (exc. insurance)	**£2,965.75**
Insurance	Group to be announced

TOTAL AS TESTED ON THE ROAD	**£2,965·75**

A dash of Köln

It's so rare that you find two variants of a model that are so good that it makes you want to sing the national anthem. And who would do it but Ford? Here we go then . . . **Deutschland über alles** . . . Well, I did mean the **European** national anthem. But seriously, though, we tried the RS1800 some while ago and were highly impressed, so when Ford brought out the RS Mexico (no relation) and RS2000, it was only natural that we should grab them. Are they indeed turd-warm? Our nitrogenous correspondent reports . . .

Whereas the RS1800's 1793cc engine is based on the old BDA and is all-aluminium and cammy and all ready to take on international treasure hunts, the RS2000 and the new Mexico units both have the Pinto-based cast-iron block with capacities of 1593cc and 1993cc non-respectively. So it's not just the bodyshell that's never been to Mexico City, and if you're after something which approximates more nearly to the old Mexico you'd better buy yourself a 1600 Sport, which will incidentally save you about £400. The RS2000 may seem a bit costly at £2857, but that's more than £60 cheaper than the RS1800. One reason for the £400 differential between the Mexico and 2000 is RS wheels; they're standard on the larger-engined car but a £117 option on the Mexico.

One other reason for the high cost of the 2000 is its Firenza-style droop snoot which, if not the styling disaster of the decade, is certainly not a thing of rare and delicate beauty. What's more, or perhaps less, the four head-lamps work roughly half as well as the two on the Mexico, Cibies or not. That at least is what other staff members tell me—I only drove the more powerful car in daylight.

Both cars were finished in a particularly bright red which was very unkind to hangovers even on a dull morning, and apart from the front end and altered stripe and badge work, the two cars are identical from the outside, sharing the squashy boot spoiler that first appeared on the RS1800. The latter is also the source of the Mexico's pop-riveted front spoiler.

Naturally, there are a lot of similarities inside, too. The very neat BMW-style switchgear, dashboard and instruments are the same except that the RS2000 has a clock; it also has a useful centre console. Both cars have what George describes as 'usual Ford quality looped acrylic carpet—looks like Tesco's cheapest'. Otherwise the trim is quite good.

The two Fords have different front seats; the 2000 has the high-backed German recliners which first appeared in the Mk I RS2000, while the Mexico has lower, non-reclining seats. I was alone among those who drove both machines in preferring the latter, but I have to agree that the smaller buckets were harder to get into, and their ridged sides could chop the unwary in half. The driving position is excellent in both cars.

The Germans certainly do as good a strengthening job on the bodies of these comparatively low volume Escorts as did AVO in the old days,

LHJ 938P

LHJ 934P

and it remains to be seen whether or not the Halewood plant is capable of as much—if indeed it ever gets the chance. The cars feel really solidly constructed and exceptionally well insulated; the 2000 in particular was notable for an almost complete lack of road noise.

The suspension of the cars is allegedly identical, with spring and damper rates raised over those of more mundane Escorts. Presumably there's a slight difference in weight distribution because of the RS2000's Silly Nose, but it can't be all that much. Admittedly my opportunities to drive the 2000 were more limited than I would have wished, but the two cars seemed to me to handle quite differently, and in this respect I favoured the Mexico.

I had read in a couple of the less reputable motoring comix that the new RS2000 had 'more understeer built into it's than the Mk I. I found that hard to believe, and even harder to believe when, with my foot hard down in third round a long smooth bend, the front end dug in and then the back came round in a sort of lurching roll oversteer. Not really too impressive. However, it seemed quite chuckable, but not nearly as much as was the Mexico. Maybe all this 'excessive understeer' is apparent on a test track, but it doesn't show up on the road. If you take a BMW 2002 round Silverstone, for instance, it's an understeering pig, while on the road it's excel-

lent. Perhaps all this proves is that you can't draw conclusions about a car from the way it behaves round a skid pan. Both cars have excellent roadholding and the Pirelli CN36HR radials give a smooth breakaway.

The ride of both machines is firm and at low speeds almost harsh, but there are no nasty clonking noises when you traverse potholes (unless of course, they're of the Hamill In Wonderland variety).

The biggest difference between the cars apart from their appearance is in their performance, and the biggest difference in this area is torque, although top end power gives the 2000 —with 110bhp—an advantage of 15bhp. The Mexico's 92lb/ft of torque doesn't compare too favourably with the RS2000's 121, and this perhaps is what led CG to describe the Mexico as 'a slug'. Which is really a bit unfair. Col even went so far as to admit that he'd had trouble keeping up with a tuff driver in his family Cortina. Well, all I can say is . . . Mmghhl . . . Snork . . . Waaah . . .

OK, so the Mexico is slow compared with the RS2000, but it certainly isn't slow slow. I reckon our test version had a faulty carb (tickover was very uneven, and response was often a bit lumpy), but even so a 0-60mph time in the high tens isn't all that bad for this class of car. It was necessary to thrash it a bit in the gears, but once you get the revs

up it moves pretty well.

There were no serious complaints from either George or Colin about the performance of the RS2000 except that Uncle George said it 'doesn't give the expected thump in the back . . .' You wanna thump in de back, huh? I give you the thump in de back, okay? He also said it 'might go well with an auto box', but he is pushing 30 . . .

I found the RS2000 a bit disappointing. Possibly we had a slow example, which is not to say that it was slow. Its acceleration is more or less the same as that of the RS1800, but you have to push it a lot harder to get the same results, and with just a single camshaft there's obviously less tuning potential. Sure it's quick, but you won't blow the doors off anything really fast —perhaps 29mpg is compensation?

Neither the 1600 nor the two-litre version liked to be trickled around town in top gear, which is what you need from an engine in town. Below 30mph they both started doing kangaroo hops, which isn't very wonderful at all.

The gearbox is common to both cars and very excellent it is. The ratios are ideal, the gate is just the right size, and fast changes come naturally. There was something very odd about the clutch of the 2000; purely a matter of adjustment, I'm sure, but there was a stickiness in the mechanism right at the engagement point which made for rather jerky starts until one got the hang of it.

I was particularly impressed by the brakes. You can go zapping down steep hills at ridiculous speeds and fly through bends barely slowing down at all in the secure knowledge that if you need to you can get down to a legal velocity in no time. There was no sign at all on either car of fade or pull, there was remarkable retardation for very little pedal effort; in fact one could describe the brakes as totally vice free, which is more than may be said of some of us.

Naturally, these cars come equipped with most of the gear which used to be considered a luxury: an excellent heating/ventilation system, a rear screen heater, and so on and I certainly reckon they'd be easy to live with as they're every bit as luxurious, ha ha, as the ordinary Escort range, only more so. Actually, they're probably quieter than the pushrod versions.

The question is, is the RS2000 worth 400 quid more than the Mexico and 800 quid more than the Escort Sport? And that's a whole different question. If you've got dollars exuding from every orifice, or daddy's buying it for you, or you can fiddle it as a company car, then the answer must be affirmative, or even yes. Otherwise you could be well advised to buy a 1600 Sport and spend a few hundred pounds on it. You may not get anything quite as well finished as the factory machine, but you'll still have a fast Escort that handles well. And a bit of cash to spare.   **PD**

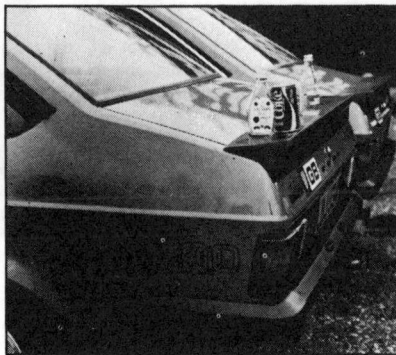

ROAD TEST

The oversleepers' cars

IT is one of the sad facts of life about road testing that the journalist tends to become terribly blase about the cars that he drives. We are often criticised by the manufacturers for being too niggly and for picking up small points which the average motorist wouldn't even notice but then it could be argued that every car should be perfect when it comes on test and that, with the pace of development of the modern car, it should show no serious faults. The result of all this nitpicking is that I rarely step out of a car feeling totally happy with its performance or handling and that I can name the cars that I have truly enjoyed driving on the fingers of one hand. I shall now have to grow more fingers, change hands and learn to count above five for Ford's new RS2000 is, in my view, one of the best cars available today.

Let's be honest; I'm not a saloon car man and it takes an exceptional saloon car to make me even vaguely interested. I have always regarded the road going Escorts as the best of a not very inspiring bunch of family/reps cars and the rally cars as being highly effective but not much fun for road use. The latest additions to the RS range put the whole situation in a different light. They set out to perform two different functions and will appeal to two different markets so it is worth looking at each car separately.

Ford would appear to have aimed the new RS2000 directly at the Dolomite Sprint/BMW320 market; by which I mean the rising young executive who wants a fast, comfortable and stylish company car. It is also aimed at the oversleepers amongst us. If, like me, you invariably oversleep, leave for all appointments at least half-an-hour late and have to cover Wales to London in a time that would certainly interest the Wiltshire Constabulary then the RS2000 is for you. It will cruise quite comfortably at 90-95 mph, accelerates from 0-60 mph in just 8.5 seconds and holds the road like a leech. In fact it is only marginally slower than the RS1800 that we tested two months ago.

For this kind of performance in almost total silence and extreme comfort you invariably have to pay and many people may find that prospect of £2857 for an Escort is something of a stumbling block. Don't be put off but compare it with £3349 for the BMW 320 or £3083 for the Dolomite Sprint, both being cars in the same class. If you are planning to do a lot of high mileage motoring then

The new Escort RS2000 is undoubtedly one of the best cars on the market today. It must represent a direct challenge to the BMW and Dolomite Sprint market.

The new Mexico has exactly the same body shell as the RS1800 though its performance feels a little disappointing. Like its bigger engined sister, it is smooth and comfortable.

The interior of the new Mexico features the standard dash panel mated to the RS console. It lacks some of the refinement of the RS2000 as does the rest of the trim. Bucket seats are standard.

you are going to have to pay this kind of money for a car. In terms of actual running costs, the RS2000 worked out reasonable but not outstanding. Despite Ford claims of over 30 mpg we never managed more than about 28 mpg over 1200 miles of varied motoring. Admittedly we didn't drive the car gently or cruise at a steady 50 mph but then I suspect that the average RS2000 buyer won't either. The car used just under a pint of oil which is again about average.

The interior of the RS2000 follows the pattern set by other cars in the RS range and is both functional and comfortable. The combined dash and centre console unit provides a positive plethora of holes for maps, matches and magazines and a clock is fitted on the passenger side, adding a touch of luxury not found on the RS1800 or the Mexico. The matt-black door trim is extended to the full height of the door and there is no painted metal band as found on the Mk. 1 version. The controls follow the Ford 'three-stalk' system, the only change being the addition of an intermittent wipe control — now standard on both the RS2000 and the Mexico.

Although the power unit remains basically the same as that fitted to the Mk. 1, there have been a few small changes which improve the torque and power. The fitting of a high-efficiency exhaust system means that the car now produces 110 bhp at 5,500 rpm and has a maximum torque of 119 lb ft at 4,000 rpm. The result is tremendous flexibility and third gear is usable from about 15 mph to about 85 mph. The close ratio box has a short throw lever fitted and is positive in the best Ford traditions though perhaps first is a little bit low compared to the other gears in the range.

The noise levels in the RS2000 have been kept ridiculously low and considering the performance, it really is blissfully quiet. Apart from the obvious points of good sound-proofing and a good body shape, there has also been modification to the bell housing and the fitting of an aluminium sump; though I suspect the latter is more to make the engine fit the Escort cross-member. The comfort is further enhanced by excellent bucket-type reclining seats which are standard fitting and considerably more comfortable than the 'Custom' seats that were fitted to the RS1800. A tall driver might find the adjustment rather restricted and everybody has difficulty getting used to the rather high

driving position but the vision is excellent and it doesn't take long to adjust. The budding racing driver who likes to be lying horizontal on the floor will just have to fit his own seat.

The car is, of course, instantly recognisable by its droop-snoot front which is merely screwed to the standard Escort front panel. A specially shaped bonnet is required and one gets the feeling that a frontal accident might prove rather expensive. Fortunately the 'snoot' is flexible to a degree and will take quite a knock before it comes to any serious harm. The twin headlamps are mounted in the front and are excellent.

So, were there any snags at all or has Ford finally produced the car that keeps the journalist happy? Well, don't try to sit your six-foot, 80-year old Grandmother in the back because she simply won't manage it. In fact don't try to sit any body in the back for any long periods since the leg-room is minimal. It is the car's only serious fault and stems from the fitting of those super-comfortable seats. Less seriously can I please have a heater that blows warm and not just hot or cold and can I please have a boot light? I knew I would find something wrong in the end.

The sporting sister

It is extremely unfair on the Mk. II Mexico to jump straight into it after a week of driving the RS2000 so, if you are going to test both cars, drive the Mexico first. The difference in performance is amazing and the Mexico is considerably harder work to drive fast. In fact, although the figures say otherwise, the new Mexico feels slower than the old one. Perhaps this is because, once again, the Mexico Mk. II is smooth and quiet whereas the Mk. I was a little rough round the edges. The new car has a claimed 0-60 mph time of 10.2 seconds and a claimed top speed of 106 mph. Although having exactly the same gearing as the RS2000 it has a lower comfortable cruising speed and we found that it was at its best at around 85 mph.

Although you could rally the RS2000, it somehow seems far too nice a car for that and the Mexico seems to be much more of the rally type car. With the engine homologated at under 1600cc (1593), it is bound to be very popular with the Gp. 1 contenders where its outstanding road-holding and handling should be at a premium. I understand that Ford have already homologated several parts for it under the new Appendix J

so expect to see a few cars out and about soon.

Apart from its obvious competition use, the new Mexico will appeal to the man who wants an Escort that's just that little bit different and is one step up from the 1600 Sport. Along with the RS2000 and the RS1800, the Mexico has a number of suspension modifications which make the ride firm and precise without being particularly hard. At the front the suspension geometry has been changed slightly while at the back the cars are fitted with twin rear radius arms instead of the standard anti-roll bar. The result is more positive axle location and consequent better handling on the loose or on uneven surfaces. The spring and damper rates have been changed and, although perhaps a bit firm for rear seat passengers, are a good compromise between handling and comfort.

The interior trim of the Mexico is standard Escort apart from the RS dash panel and it lacks the centre-console and refinements of the RS2000. The seats as standard are of the bucket type with fixed backs and proved to be a little hard for long distance travel. Nevertheless the driving position is comfortable and relaxing and the Mexico seemed to swallow the M4 with much the same ease as the RS2000. The one advantage of the seats is that there is now room in the back for passengers and the car is once more a true four-seater.

The power unit of the new Mexico is lifted almost directly from the Capri and Cortina range and the 1600 unit uses the same block as the more powerful 2-litre. The difference comes about through the bore and stroke, the Mexico measuring up at 87.65 mm x 66.0 mm. With a revised exhaust system again, the Mexico engine produces 95 bhp at 5,750 rpm and 92 lb ft of torque at 4,000 rpm and the problem seems to be that the power doesn't come in until about 4,000 rpm. Thus, in order to get the car to move along smartly you have to be constantly playing about with the gearlever. That shouldn't worry the enthusiast but it could prove to be disconcerting to less discerning motorists.

The body of the Mexico certainly attracts attention as does the RS2000 and our particular car was finished in white with black stripes. Very fetching but not very practical at this time of year. Alloy wheels are an optional extra and the standard tyre fitting is the new steel Pirelli Cinturato. It appears to be man enough for the job though I found that it felt a little uncertain on slippery surfaces. The handling more than compensates though and getting any of the RS range out of line whilst driving on the road would tend to suggest a leaning towards the foolhardy rather than the safe.

Priced, as it is, at £2443.50 the Mexico falls neatly into the same bracket as the Princess 1800 HL (£2398), the Audi 80 LS (£2468) and the Fiat 132 GLS 1600 (£2398). It is certainly the most sporting of that bunch but whether it would match them in trim I would doubt. The Mexico is a sporting Escort, it will appeal to the sporting driver and, if he can't afford an RS2000, then he should be well pleased with his choice.

"This weekend of you RS

— where did you go?"

by Michael Scarlett

"WELL REALLY, JUST CORN-WALL and back on Saturday, and Perthshire on Sunday."

"Silly."

"Hm." It was silly really, by some people's standards. The car I was using, Ford's Escort RS2000, would make a mockery of the 1,680-odd miles we were to drive by Tuesday evening. Lots of drivers have driven further in less time, other infinitely harder routes; nothing news-worthy in doing the MCC's 55th Land's End Trial with your father as crew returning to Richmond (Surrey) the same day, driving the Escort with a roof-rack and boot-full of luggage to rejoin your wife in Scotland on Sunday, spending Monday and Tuesday morning in the open air and driving back south again in the afternoon. But it did say a little for the versatility of a good, modern, and apart from the funny nose and the rubber thing on the bootlid, unpretentious car.

Having asked for an early number, we were a little em-barrassed by the conspicuous-ness of being the first car to start

panzees' Tea Party. Sidecar outfits normally look as unstable as they are; what other vehicle has survived from the primitive days of locomotion which turns if you let go of the controls? Trials chairs with narrow track, axle-high ground clearance and platform on which the passenger perches cling-ing to a strap, his knees seemingly by his ears, his head as high as the driver's — they made thge wonder-fully relaxed comfort of the Es-cort's rally-style seats seem the utter depths of sybaritic depravity — especially over the 400-odd miles of the Land's End. We had head-lamps, too, superb halogen ones; their minimum lighting looked like candles in comparison.

Mist rolled over us after mid-night, as that red moon had warned. At Sparkford, the first stop and the first, very little rest (it might have been awkward if wet) we met Steady again, who explained that he was now going to Weymouth, because the mist pre-vented him getting to Southamp-ton. What was he driving? He was

Opposite parts of the world; left, the RS2000 in M.C.C. Land's End Trial trim swinging round a tight corner on the only hard-surfaced hill, Ruses Mill; above, pausing on the road into the Sma' Glen — a splendid trials car and a delightful fast touring car.

from the "London" depart, at Fleet service area on M3 (10.28 on Good Friday evening). I forgot that when at the last minute my old colleague Ronald Barker who during our expressions of mutual surprise — "Well, I'm blowed, amazing to see you here Steady, why aren't you taking part?" — explained that he was on his way to Southampton, to France.

We ambled off, marvelling at the ways of coincidence. I am always fascinated by Stonehenge; the dark would hide it from us in spite of a reddish, slightly bitten cheese of a moon rising around 11.30. But the loom of headlamps momentarily beyond silhouetted it sinisterly, even though it was on our right. We caught up with two trials sidecar outfits swaying round each bend like, without wishing any disre-spect to the bravest and most hardy of fellow competitors, two pairs of performing monkeys escaping on three wheels from the Cham-

"riding side-saddle on a Yamaha", presumably to make room for the two companions he seemed to have with him. It wasn't until the next test, Sug Lane, a dry narrow little track which disappeared up into the darkness at the side of a lane; winding between the two narrower walls of a bridge, that we learnt the truth, that he was crewing for Tom Threlfall in the latter's Model A Ford.

What was virtually fog slowed everybody. We followed some Imps. King's one amused us at each corner, only the right stop lamp lighting up, while the left rear lamp went out, as though the car were winking — "No need to slow for this one really." The air cleared near Taunton around 4 am though we were still very grateful for the interruption-wiper setting on the Escort; ideal for British weather The next hill, Edbrook, was fun, quite steep, and twisting, Father in the back seat hopping from side to

CONTINUED ON PAGE 97

IT'S BEEN A FACT FOR YEARS that if a passionate driver wanted a car capable of doing all the things sports cars are meant to then he'd have to forget about the few soft-tops left on the market and opt instead for one of the worked-over tintops. For the traditional sports car — in Britain, anyway — the crunch arrived with the Mini Cooper, and who can't remember the tales told in pubs by smug minimen fresh in from fixing up some string-backed type in his MGB? But it was Ford's answer to the Mini Cooper that really put the seal on it. Along came the RS1600, and Ford almost had to build a warehouse to store all the trophies it won. Any sports car driver who knew of its capabilities had to slink away and hide, and the boy racers had a new idol. The RS1800 took over when the new Escorts arrived, and so far there's been nothing to match them as production racing and rally cars. Vauxhall's Firenza came and went; the Lancia Stratos costs as much as a Ferrari, and you can't go to the corner dealer and order one even if you do have the money. The closest challenge to the Escort, so far as the man in the street is concerned, now comes from General Motors with the Opel Kadett GT/E. It isn't yet known if Fiat's weapon, the 131 Rallye, will come to Britain, and it will be a while before Vauxhall's next try, the rally-pack Chevette, reaches the showrooms.

Now, for serious, big-league rallying or racing you're going to need the out-and-out competition versions if you're to have any hope of getting close to the big names. In the Escort, that means 275horsepower on the track and 230bhp for rallies. What we want to ascertain here is whether the production RS1800 sitting in the Ford Rallye Sport dealer's window with 125bhp under its bonnet can serve satisfactorily as an everyday road car for some one who wants a sporting saloon capable of putting up a reasonable performance in club racing or rallying at weekends. The second point to be considered is whether or not the Opel Kadett GT/E might do the job better. For a start, you're not going to get either car cheaply so the boy racers are fairly effectively excluded. The 'basic' RS1800 costs £3206 and the Custom — which comes with fully-trimmed doors and rear quarters, reclining seats with head restraints, clock, centre console, glove box, carpeted boot and inertia reel seat belts — tested here is £3342. The Opel manages to be cheaper than the basic RS1800 at £3166. Remember, like the Opel, the Escort RS models are imported from Germany now that FAVO has closed down in Britain. It is also pertinent to remember that in this price range you're in the realm of Alfettas, Lancia Beta 2.0litres, saloons which have more than a fair measure of sporting character within their much more sophisticated natures and capacious bodies.

GIANT TEST: FORD ESCORT RS 1800/

STYLING, ENGINEERING

Ford leave the pretty nose to the RS2000; the serious car gets a serious nose with a deep spoiler and another one (of soft rubber) adorns the bootlid. They're required, Ford say, primarily for circuit racing. Dimensionally, the RS1800 body is unchanged from the standard car, but it has been strengthened in all the stress areas to withstand the rigours of competition. It's tough, it looks tough; and apart from the two-tone stripes along the sides it has no boy-racer pretentiousness.

Does the Opel? The shape is clean — a pure coupe compared with the Escort's two-door saloon lines — and its front spoiler is an integral part of the body design of

all Kadetts. But the GT/E comes with what many people consider to be a garish and tasteless paint-job: black up to the waistline and bright yellow for the rest, with huge GT/E lettering on the mudguards. So what the Opel gains on body shape it loses in the paintwork. Like the Escort, it is strengthened in vulnerable areas.

Most interesting technical aspect of the Escort is its engine. A direct descendant of the original Cosworth FVA racing unit, the RS1800 BDA grew from the old 1600 through an increase in bore from 80.97mm to 86.8mm. The stroke remains unchanged at 77.62mm. As before, the cylinder block is aluminium, and so is the four-valve

per cylinder head and its twin cams. But in place of the twin 40mm side draught Webers that used to adorn the RS1600, the new engine has just one twin choke unit — a move designed to provide a better power/economy compromise and better flexibility in traffic. We shall see just how well this works a little later on.

From its 1854cc capacity and a 10 to one compression ratio, the BDA develops a stirring 125DINbhp at 6500rpm, and backs that up with the pull of 120lb/ft of torque at 4000rpm, a fairly low peak for a competition-oriented engine. Such figures suggest a great deal of well-considered work on Ford's behalf, and that the Escort will be rather

Richard Davies

very much stiffer than in normal Escorts and the anti-roll bar is heftier. Steering is rack and pinion, with 3.5 turns lock to lock.

The Opel's live axle is located by trailing links, a Panhard rod and an anti-roll bar that the other Kadetts don't get. A torque tube provides further help in keeping things under control, and at the front there are upper and lower wishbones with anti-dive geometry, an anti-roll bar. As in the Ford, the shock absorbers are Bilsteins. Both cars also run on 5.5 x by 13in rims shod with 175/70HR13 radials — Pirelli CN36 SM's in both instances in this test. Six-inch-wide alloy wheels are optional on both cars. Steering in the Opel is rack and pinion, with 3.5 turns lock to lock.

PERFORMANCE

These cars exist for performance, and for handling. Remove either faculty and they lose their validity as well as their appeal. But it takes only a few moments in the Escort to know that it has performance a-plenty, and that it is accompanied by a great deal of appeal.

The BDA engine starts easily — there's an automatic choke and is fuss-free from the outset. Gone is the all-or-nothing characteristic of the old RS1600 where you went nowhere until the tachometer reached 4000rpm and you then disappeared towards the horizon in a cloud of tyre smoke and scrambling pedestrians. This one has excellent flexibility and a very smooth progression up through the rev range. Certainly, there is a noticeable increase in urge beyond 3500rpm, and the top end bite is quite outstanding; but living with this car in suburban streets or heavy City traffic is no problem at all. Indeed, it's rather like driving an Alfa Romeo or a twin cam Fiat. It will pull without grumble from as low as 1500rpm in top.

And yet, so great is the appeal of the engine that it's almost impossible to resist using it as it should be. You find yourself changing down to the very usable first gear for tight corners or for traffic lights, much as you would for second in other cars. You use all four gears here, rather than only the top three (or four, if you're that lucky). On the way up, if you throw the throttle right open, you'll reap the benefit of having a very respectable 16.12lb/bhp power-to-weight ratio. The little car rockets forward with the engine just getting stronger and stronger the more it revs (the upper limit is 7000) and it whips to 60mph in 8.6 secs. Put your foot down at 80mph (4300rpm)) and it surges ahead with real eagerness. You can feel the strength in the engine. Even upwards of 100mph there is plenty of urge left, and the car will run on to 118mph. Quite apart from the engine's sheer performance, it has such a delect-able feel about the way it works, and even with the single carburettor it makes all the right noises. Sports car owners of the old school should be so lucky!

OPEL KADETT GT/E

nice on the road. The competition Escorts come with all-synchromesh five-speed gearboxes, racing clutches and limited slip differentials; the road-going RS1800 makes do with a four-speed transmission and an ordinary differential (3.54 to one) that provides 18.5mph/1000rpm.

Opel have taken a rather different approach with the GT/E's powerplant in a shoe-horning approach that corresponds more with what Ford did when they created the RS2000 than with the RS1800. This means that the Kadett has been given the 1.9litre fuel-injected engine from its big sister the Manta GT/E. Compared with the Escort, and apart from the Bosch L-

Jetronic injection, it's a rather crude engine: iron block and cylinder head, and one cam (mounted in the head, in typical Opel fashion to work short pushrods and rockers). The Opel engine is considerably more oversquare than the BDA, with a 93mm bore and 69.8mm stroke, dimensions that provide it with a capacity of 1897cc. It runs on a relatively modest 9.2 to one compression ratio, and at 5400rpm it delivers 105horsepower. With a fairly flat torque curve, you get 110.7lb/ft between 3400 and 4600rpm. Although a four-speed transmission is standard in the Opel, our test car had the optional five-speed gearbox coupled to the standard 3.44 to one differential. A

3.67 to one limited slip differential is optiona..

All Opels are prime examples of just what can be done with straight-forward suspension engineering. In this case, however, it is perhaps the Ford that is the most impressive, for it begins life with a rear suspension even more basic than the Opel's coil-sprung live axle: the Escort has leaf springs. But good development and extra location by radius arms means that the axle is kept firmly in place and that spring wind-up is virtually eliminated. Naturally, the spring rates are very firm and they are supported by Bilstein gas-oil shock absorbers. At the front, Bilstein cores are used in the MacPherson struts, the springs are

The Opel has plenty of get-up-and-go, but it has neither the outright acceleration of the Escort nor its under-bonnet appeal. It too is torquey and flexible, it too will rev hard but somehow it just feels rather ordinary; it does not have the delicious and unmistakable sporting character of the Escort. Only twin cams can give you that.

And although the Opel appears to be faster than the Escort, it proves to be rather slower, taking 9.8secs to reach 60mph. It is only 3lb heavier, but it is 20bhp down so the power-to-weight ratio is 19.2lb/bhp However, with a five-speed transmission in which top is direct and all the ratios beneath it are extremely close, there is plenty of in-the-gears urge, and a foot flattened at 100mph in the Opel brings instant and pleasing results, and a top speed of 115mph.

Because the five-speed transmission's ratios are so close, it takes a while to get used to the Opel, for it can seem fiddly at first, even though the engine itself is in no way fussy. It's just that you seem to need to change gears so often — the car is forcing its aggressive little character upon you. You'll like it best when you can find a good few miles of fast, fairly bendy road where you can keep your foot down and work up through one gear after the other, quick as flash, the tachometer dropping only around 1000rpm after each upward shift.

Driven as hard as this, the Opel returns 24mpg. Take it fairly quietly and it will provide you with an impressive 32mpg. The Escort is more economical when driven hard — at 25mpg — but usually only runs to a shade over 30mpg at the other end of the scale. Considering their performance, both cars are pleasingly frugal.

HANDLING AND ROADHOLDING
As we have seen, then, both these cars deliver the goods so far as performance goes. And nor do they falter in this other vital area. They have prodigious grip at all times, and they have the sort of handling to make any keen driver just want to keep on going for as long as possible. Normal sealed-road bends present no problems; both cars just hang on for dear life and point obediently in the direction they're told to follow — or would *asked* be a better word? They have balance, and they have the power to let the driver exploit that balance. Back off a little coming into the bend and any understeer is hurled away; power-on from the apex onwards and you can oversteer at well, and always with the enjoyment and security that stem from outstanding controllability. But while their cornering capabilities are equally-matched, so far as we can tell, there are differences in the way the two cars behave.

The Ford, like its engine, feels tauter and more responsive from the outset; more alive. With a good deal of negative camber at the front wheels and plenty of castor, the steering has a very meaty feel (although it's not heavy). Backing

this up is a chassis that feels as lithe as we've encountered. The car feels small and very tidy beneath you— the old glove simile really does apply here. You grasp the wheel in your hands and you are in no doubt that this car means business, that's it's honed razor-sharp.

The Opel doesn't have the same sort of lean, bunched-muscle feel about it. It feels and acts more like a big car, with lighter steering and a more fluid sort of nature to it as it gets along the road. On normal B roads, the Opel displays certain advantages over the Escort. It retains its directional stability better over crests taken at high speed whereas the Escort squirms a litle, and there are instances when the extra gear means that it can get out of bends with more oomph than the Escort can. It is here that the Opel feels a bit smoother — that big car impression again — than the Escort.

On the dirt — ah, that's another matter. Despite its better axle location, the Opel doesn't seem to get its power down as well as the Escort does and it is not as stable at high speed. The Kadett requires a very smooth, fluid driving style and absolute competance if it is to give its best. So driven, it is very fast and can be swept through loose-surface bends at extremely high speeds. The Escort immediately feels easier to drive on the dirt than the Opel because of its tautness, its feeling of unity. All its movements are well-controlled, and you feel that you can do anything with it, mostly because you know that it is so responsive. And yet it seems to allow you enough time to take the right action as the tail comes out. When it gets beyond the driver — perhaps in over-correction after being unbelieveably sideways — it is still retrievable after the point where, in the Opel, you'd have stopped bothering and let it spin. Just snap the Escort's wheel back the other way, wait for the right moment and then tread down hard on the power again and it comes straight. It is quite, quite outstanding. More importantly, in everyday road-going situations it is magnificently safe; so is the Opel.

COMFORT
The pleasant thing about these two cars is that even with such crisp handling and the fact that they are realistic bases for competition cars, they still manage to ride fairly well. Naturally, their rides are very firm; but they are not uncomfortable. Indeed, during our test both cars were driven several times over roads that a week earlier we'd travelled in a Datsun Violet, which has no sporting retensions and

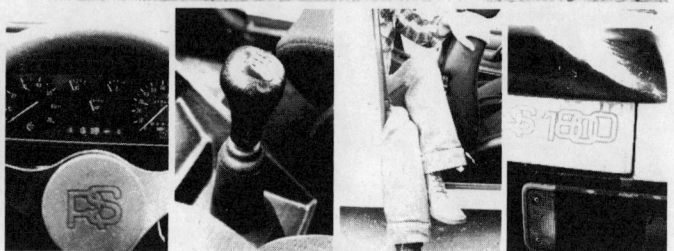

FORD ESCORT RS 1800

ONO 804P

Trim and taut Escort handles magnificently at all times (top). Above, from left: driving position and instruments are outstandingly clear, stubby gearshift is superb but access to and from rear is awful; dirt and mud collects on the rear panel. Special rally-type buckets are extremely comfortable (left), the little twin cam BDA engine is a pure gem (below) and the carpeted boot is plenty roomy (bottom)

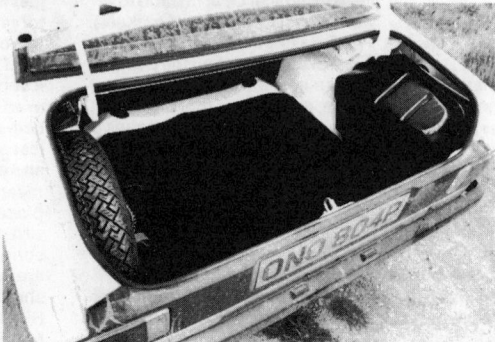

certainly does not handle especially well. Yet the rally cars' rides were immeasurably better; indeed, the faster they go the better they ride. Inherent with the firmness is that ever-pleasant feeling of very well-controlled damping, and even though the suspension travel is quite short they rarely touch the bumpstops. Again, the tremendous tautness of the Escort is impressive; it has a real feeling of quality about it. Although for the most part the Escort's ride is more jiggly than the Opel's, there are times when the Kadett's rear axle is caught out and sort of bounces underneath the car for an instant or two.

It's with the seating that the Escort really scores. It has two fully-bolstered rally-style buckets that offer driver and passenger excellent comfort as well as locating them perfectly. The only problem is that they steal a lot of what little rear legroom there is in the Escort. The Opel's seats are not nearly so good; the squabs are too flat and lack real lumbar support in a car where the driver is often hard at work. However, the Opel has both more rear room and a better shaped rear seat and can serve as a four-seater more satisfactorily than the RS1800. Strangely — for it has a viscously-coupled fan — the Kadett has a lot of fan thrash, and suffers from pronounced tappet noise once past 4000rpm too. This makes it noisier than the Escort, even though it transmits less road noise. At 80mph the Escort engine has a boom period, but after that it is very quiet, and relatively relaxing to drive. The Opel is rather noisier and becomes tiring on long hauls. Boots are good in both cars; finish is better in the Escort, and so is equipment.

DRIVER APPEAL
Well, obviously yes. Full instrumentation is a necessary feature in both cars, but the instruments are arranged better in the Escort. The new Ford instrument panel is as good as they come: speedo, tacho, oil pressure, temperature and fuel gauges are all directly in front of the driver and their white-on-black faces are easy to read. There are no reflections. The Opel's gauges are clear, but the temperature, oil pressure and fuel dials are mounted separately, low-down in the centre of the dash.

At first, the Escort's driving position might seem a bit sit-up-and-beg. But the moment you hit the dirt and start to flail about in opposite lock you discover that it's absolutely perfect. The Opel's is equally good; pedals are well-located in both and so are the gearshifts. But whereas the Ford's shift is outstandingly good, the Kadett's requires some familiarity because the movements are very close together and it can be all too easy to wrong-slot. Our test car's shift was playing up too; it was extremely difficult to engage reverse (the pattern is the old race-car pattern with first down to the right and the top four in the H). Vision is best in the Escort, but the Opel is still good; the narrowness of the cars makes them so easy to place in at speed in tight conditions.

SAFETY
With potent brakes to back up pin-sharp handling, tenacious roadholding and excellent stability, these cars are that much harder to involve in an accident, provided the driver is not stupid, than lesser machinery. Such good seats help in the Escort, both with staving off long distance fatigue and with location for maximum driver control. The Ford's halogen lights are commendably strong, but while the performance of the square Opel units is good they require conversion to rhd. Both cars have laminated windscreens, good seat belts and strong bodies.

Ford Rallye Sport and DOT dealers can supply the obligatory

rollcages and fire extinguishing systems — as well as sump shields — for those going rallying.

CONCLUSIONS
There is no doubt that the Escort RS1800 is not only perfectly acceptable as a very sporting, everyday road car; indeed, it is a desirable one. So long as one is seeking maximum fun per mile. Its performance might be taken for granted before one even slips behind its wheel. So might its handling. What will surprise many people is its relative sophistication and comfort, and indeed its appealing character. Most of that appeal comes from the engine: it is flexible, smooth and delightfully strong, especially at the top end. It has real bite all the way to 7000rpm. The gearchange is a further delight, and the handling backs it all up. The seats round it off, and that the car is not too noisy is another unexpected surprise.

Good as the Opel is — and we'd be quite happy to have one, thank you— we think the Escort does the job better. It is more readily available, is sold rhd (although lhd can be an advantage in narrow road, and is no real hardship to most drivers), it has a more appealing engine that provides greater performance (the optional diff would take the Opel closer to it in acceleration) and, on the dirt at least, we believe it has a handling advantage. Moreover, for the owner who will enter it in rallies or races, there is such a wealth of experience and parts available through the Ford network and a host of smaller garages, although Tony Fall's DOT operation is coming on strong with Opel facilities and could build a GT/E in rhd if you really wanted it that way.

Kadett rolls little, handles well but isn't as easy to control as the Escort. Above, from left: thick-rimmed wheel is good to use, rear accommodation is better than in the Ford but the instrumentation is clumsy and ventilation inadequate; lights are still set for lhd and the Opel doesn't get its power down as well as the Escort. Tartan interior (left) is garish and seats are too flat. Injection gives the 1.9litre engine its power (below) and the boot (bottom) is surprisingly deep

FORD ESCORT RS 1800

Capacity (cc)	1845
Bore (mm)	86.75
Stroke (mm)	77.62
Compression	10.0 to one
Valve gear	DOHC 16-valve
Carburettor	twin-choke Weber
Power (DIN/rpm)	125/6500
Torque (DIN/rpm)	120/4000

TRANSMISSION

Type	Four speeds, all synchro
Ratios — mph/1000rpm	
1st	3.37 to one/5.5
2nd	1.81 to one/10.3
3rd	1.26 to one/14.8
4th	1.00 to one/18.6
Final drive ratio	3.54 to one

CHASSIS & BODY

Type & method of construction	Unitary
Suspension front	Macpherson struts, lower links, anti-roll bar
Suspension rear	Live axle, leaf springs, radius arms
Steering type	Rack and pinion
Turns, lock to lock	3.3
Turning circle	32.8ft
Wheels	5.5J x 13
Brakes, type	Discs and drums

DIMENSIONS (inches)

Wheelbase	94.5
Track, front	52
Track, rear	51.7
Length, overall	156.6
Width, overall	62.8
Ground clearance	6
Fuel tank capacity	9gals

CABIN DIMENSIONS (inches)

Headroom, front	37.8
Legroom, front (seat back'	39.3
Headroom, rear	36.9
Shoulder room, front	49.7
Shoulder room, rear	49.7
Luggage capacity (cu. ft)	10.3

MAINTENANCE

Total cost 12,000 service	NA
Sump (capacity/oil grade)	6.7 pints 20/50
Oil change intervals	3000 miles
Grease points/intervals	None
Time for renewing exhaust system	
Number of UK dealers	71 Rallye Sport

MECHANICAL SPARES PRICES

Engine, new	1282.75

BODY PART PRICES

Front door (primer)	£10.95
Front bumper	£12.17
Bonnet (primer)	£29.63
Windscreen	£26.04
Headlamp unit (each)	£8.71
Grille	£14.33

TOTAL COST, INCLUDING CAR TAX AND VAT

Price without extras	£3342
Price as tested	£3342
Model range price span	£3206-3342

GUARANTEE

Length and conditions	12 months/unlimited mileage

OPEL KADETT GT/E

Capacity (cc)	1897
Bore (mm)	93
Stroke (mm)	69.8
Compression	9.2
Valve gear	OHV, cam-in-head
Carburettor	Bosch fuel injection
Power (DIN/rpm)	105/5400
Torque (DIN/rpm)	114/4200

TRANSMISSION

Type	Five speeds, all-synchro
Ratios — mph/1000rpm	
1st	3.875 to one/4.8
2nd	2.399 to one/17.75
3rd	1.763 to one/10.55
4th	1.259 to one/14.77
5th	1.00 to one/18.59
Final drive ratio	3.44 to one

CHASSIS & BODY

Type & method of construction	Unitary
Suspension front	Wish bones, coils, anti roll bar
Suspension rear	Live axle, coils, anti roll bar, trailing arms, Panhard rod
Steering, type	Rack and pinion
Turns, lock to lock	3.5
Turning circle	32 ft 7in
Wheels	5.5J x 13
Brakes, type	Discs and drums

DIMENSIONS (Inches)

Wheelbase	94.3
Track, front	51.3
Track, rear	51.2
Length, overall	162.4
Width, overall	62.2
Ground clearance	6.8
Fuel tank capacity	9.4 gals

CABIN DIMENSIONS (inches)

Headroom, front	37.4
Legroom, front (seats forward/back)	42/47.1
Headroom, rear	38.7
Legroom, rear (seat forward/back)	23.6/28.6
Shoulder room, front	50.7
Shoulder room, rear	50.4
Luggage capacity (cu. ft)	10.9

MAINTENANCE

Total cost 12,000 service	2.7 hours
Sump (capacity/oil grade)	6.7 pints 20/50
Oil change intervals	6000 miles
Grease points/intervals	None
Time for removing/replacing engine/gearbox	N/A as yet
Time for replacing clutch	N/A
Time for renewing front brake pads/shoes	N/A
Time for renewing exhaust system	N/A
Number of UK dealers	181

MECHANICAL SPARES PRICES

Engine on exchange	Short motor £183.60
Gearbox on exchange	£119.88
Differential on exchange	£99.90
Clutch unit	£22.84
Brake disc	£15.44
Set brake pads	£8.80
Set drum linings	£6.58
Fuel pump	£11.01
Damper (front)	£5.14
Exhaust system	£77.81
Oil filter	£1.18
Alternator exchange	£39.58
Starter motor exchange	£33.91
Speedometer	£14.09

BODY PART PRICES

Front door (primer)	£56.16
Front bumper	£19.27
Bonnet (primer)	£41.20
Windscreen	£27.75
Headlamp unit (each)	£29.10
Grille	£11.34

TOTAL COST, INCLUDING CAR TAX AND VAT

Price without extras	£3166
Price as tested	£3166
Model range price span	£3166

GUARANTEE

Length and conditions	12 months unlimited mileage

```
ACCELERATION
```

	0-30	0-40	0-50	0-60	0-70
Ford	3.0	4.6	6.5	8.7	11.7
Opel	3.3	5.3	7.8	9.8	14.7

```
SPEEDS IN GEARS
```

	FIRST	SECOND	THIRD	FOURTH	FIFTH
Ford	0-35	0-66	10-94	15-118	
Opel	0-31	0-50	8-68	15-96	25-115

FUEL CONSUMPTION: Ford 25-30mpg, Opel 24-32mpg

RunningReport

FORD ESCORT SPORT 1600

At over 30,000 miles our Sport is beginning to show its age yet has still to let us down. Gordon Bruce reports

MY COLLEAGUES would really be better qualified to bring HYL's story up to date than I, as for the last month or so it has acted as saviour to those otherwise deprived of wheels and thus covered relatively few miles in my hands. Spotting it intact on our roof car park after the Christmas break I could but hope it was still its old reliable self. In that it had not stranded anybody in the midst of our frozen country it was still going okay, but if the truth were known all is not quite well and now having topped the 30,000 mile mark our Sport is beginning to show its age.

Problem one was first noticed some weeks ago when having

The exhaust pipe doesn't look quite like this at the moment but thanks to some demon wiring by P. Dron should last a little longer

scraped the frost from my windscreen I was in the process of reversing back from the space outside my house and found to my horror that the gear-lever showed a marked reluctance to move from its reverse slot. At this stage I suspected that the relevant gear was actually binding on its shaft, but more recently the phenomenon has been accompanied by a tendency for the car to continue inching backwards even with the clutch fully depressed, thus showing the fault to lie within the clutch rather than the box itself. The tendency for the clutch to judder as well makes me think it now has little life left in it. Fortunately the gear change problem is confined to the first change of the morning and thus so far has not been particularly inconvenient.

One or two odd noises have found their way into various corners of the machine, for instance what I take to be a dry fan bearing causes the heater blower to sound like a squadron of Spitfires for the first few minutes of running — afterwhich, like the clutch, it recovers to work normally. The other mutterings come from the axle which now hums away happily to itself all the time. There is a degree of backlash in the differential as well. However, knowing Ford diffs of old, I suspect it too will run on for a good few miles yet without causing any real drama. The exhaust on the other hand will not

stand the test of time. At present it lives thanks to a judicious piece of wiring by P. Dron who somehow managed to rejoin the expansion box with its flailing tail pipe; the ravages of rust having caught up in what is now nearly two years of running.

The final clue to what has been quite a hard life is what I take to be either unevenly worn discs or oval drums. Whatever the cause gentle deceleration cannot be accomplished without consistent grabbing. The pedal is also long in travel, suggesting the need for adjustment despite a service that occurred only 2500 miles ago.

Having said all that, the old Sport still performs really rather well and if driven in suitably exuberant fashion can surprise the owners of noticeably bigger-engined machinery — certainly there is little or nothing wrong with its venerable 'Kent' engine. It starts promptly, performs better than ever and continues to return in excess of 30 mpg. It is surprisingly quiet if the revs are kept below 5500 rpm (above which there is very little power anyway) and what is more it requires no oil between services.

Perhaps the most exciting thing to happen to HYL since we last wrote about it is the locking of the keys in

its boot by the Sports Editor. In a hurry to head north on a Friday eve Mike apparently felt there was no time to speak to his friendly Ford dealer about trying various spares to determine the number (needless to say the little tag with the record of the number was in the boot with the key) so set about the rear end of the car with the help of, and I quote, "his girl friend, a mechanic, and a large policeman." I might add that the boot catch and the rear seat both showed evidence of the strong arm(s) of the law but I'm happy to say the keys were finally persuaded through the little holes in the rear bulkhead and Dood was able to continue on his way.

A non-mechanical failure but a nonetheless disappointing occurence is the sudden and unexpected death of the radio. P. Dron and I were traversing the wilds of Box Hill one morning when without warning Capital Radio was cut-off in its prime and, in my car at least, has never come back on the air. Having an hour to kill one day (though the first to admit that I know absolutely nothing about electrical devices) I released the offending appliance from the facia and stripped it as far as I dared. Sadly, I'm none the wiser, and what was a good radio still doesn't function.

Rumours about our Escort's replacement are now rife within the office and this could well be my penultimate report on HYL. Whether it is or not, I feel it worth emphasising that I am still enjoying the car and feel what faults are beginning to manifest themselves would have taken longer to materilise in the hands of one private owner. I would have no hesitation in spending my own money on an Escort if I were in the market for a car in that class.

IN BRIEF

Model: Ford Escort Sport 1600
When bought: March 1975
Total mileage: 30,165
Price when bought: £1860
Price now: £2393
Value now: £1825
Overall mpg: 30.8 mpg

Faults and failures

0-6000: Corrosion on delivery on one bumper bracket, plus loose head restraint collar and boot lock. Pronounced running-on and rattling rear suspension mountings. Gearchange (second to third) often heavy.

6000-8000: Running-on, initially cured, now returning, gearchange still reluctant. Petrol tank collapsed under pressure due to blocked breather.

8000-12,000: Exhaust manifold developed a leak and was changed; parcel shelf collapsed; several rattles around facia.

12,000-18,000: Running-on now worse than ever, erratic idle and constant threat of stalling. Crypton tune at Ford dealers only partly cured problems, still runs on. Parcel shelf collapsed again. Finally halted when carburetter jet fell off. Carb and timing fixed promptly but still runs on although it now accelerates well. Front brake pads replaced.

18,000-24,000: Occasional running-on and misfire from an engine that seems permanently down on power and revs. Wiper blades replaced.

24,000-30,000: Broken exhaust tail pipe. Clutch sticking and judder. Noisy heater fan through over-dry bearing. Wiper blades changed again. Grabbing brakes. Radio ceased to function. Noisy axle.

ESCORT CREATES A DUST STORM

"THE SILVER JUBILEE SAFARI Rally could prove to be the fastest and toughest Safari ever", said Ford Competitions Manager Peter Ashcroft on his return from a recent visit to Kenya.

While two of Ford's Silver Jubilee Safari entries - Britain's Roger Clark and Finland's Ari Vatanen - were recording two more international rally victories for the Escort in Europe, Bjorn Waldegaard of Sweden in company with co-driver Hans Throzzelius and three engineers led by Peter Ashcroft spent two weeks in Kenya test driving a Ford Escort RS1800 under safari conditions.

The ten-day Escort test programme culminated in Bjorn Waldegaard competing unofficially at Number 'Zero' in a local Kenyan event the 1200 kilometres Lake Rally.

This 15 hour rally is run from 1900 hours until 1000 hours the following day so as with the International Safari, many of the rougher sections of the route are completed in darkness.

Ahead of the field and just four hours from the start Bjorn encountered one of the many hazards which make the Safari Rally the toughest and most demanding event in the international calendar.

A mud hole one metre deep, 6 metres wide and 60 metres long completely blocked the rally route.

Naturally it was unmarked and Waldegaard's Escort in company with three other leading competitors including local Kenyan driver Vic Preston Junior - the fourth member of the Ford works Escort Silver Jubilee Safari team - plunged into the hazard at more than 130 kph. The Escort came to a stop within 10 metres and it took one hour's hard work by the crew to dig themselves out.

Determined driving by Waldegaard regained the time lost and before the first section of the route had been completed, Bjorn was again running in an unofficial first place. The test run came to a final conclusion when a hidden boulder smashed the rear axle differential 150 kilometres from the end of the rally.

The test Escort RS1800 will now remain in Kenya until after the Silver Jubilee Safari. It will be used by other members of the Ford team for their reconnaissance drives.

"The Safari always presents particular problems," says Peter Ashcroft. "The 6000 kilometre route runs from sea level to above 3000 metres. At high altitudes engines lose power so we have to revise our carburettor settings to produce the best performance possible at all altitudes".

The 2 litre 16 valve twin overhead camshaft engines in the Safari Rally Escorts will be tuned to develop 240 bhp at 8000 rpm. Smaller valves than those used for European events will improve the engines torque and the compression ratio will also be reduced.

Softer suspension settings with increased suspension travel will be built into the works Escort RS1800s and the tyres will be steel braced radial ply Dunlop tyres with a mud and snow tread pattern.

Two sizes will be available to the Escort drivers in the Silver Jubilee Safari - a 175 x 13 on a 6 inch width rim and a 195 x 13 on a 7 inch width rim.

Such is the severity of the Safari route that at rally speeds a complete set of tyres will not last more than 450 kilometres.

An underbody shield will also be fitted to protect the engine and transmission but the general mechanical specification of the cars will be identical to the European rally car.

Lighting regulations in East Africa allow greater freedom for the positioning of auxillary lamps. The Safari Escorts are being fitted with special brackets which are mounted on to the tops of the front fenders close to the windscreen pillars. These brackets can, if necessary, carry additional halogen spot lamps.

Extra powerful air horns will also be fitted to all the works Escorts so that the drivers can give advance warning to animals.

"In addition to our concentrated test programme we have received invaluable assistance from our local crew of Vic Preston Junior and John Lyall", says Peter Ashcroft.

In 1976 Annabels - London's most exclusive night spot - entered an Escort RS1800 in the Safari for Robin Hillyar who drove his Ford Taunus to outright victory in 1969. At the end of the 1976 Safari the car was taken over by local driver and son of a former Safari winner Vic Preston Junior, who has driven it with great success in the East African Rally Championship winning three events outright.

All the experience which he and co-driver John Lyall have gained in 12 months of East African competition has been transmitted to Ford in Europe to assist in the development of the Escort and the preparation of the cars for the Silver Jubilee Safari.

"Junior Preston's assistance for the Safari has been invaluable", says Ashcroft "the experience and information which he has provided enabled Bjorn to cover certain sections of this year's route at speeds 30 per cent higher than are required for the Rally".

Flat out over one of the sections of this year's Safari Rally Bjorn Waldegaard of Sweden in his Ford Escort becomes completely airborne.

Bjorn Waldegaard and Hans Thorszelius of Sweden talk over suspension settings and choice of tyres with Ford engineers during a short break in their ten day test programme. Extreme left is Ford Competitions Manager Peter Ashcroft.

AutoTEST

Ford Escort 1600 Sport

**Fast and sporting version of highly-popular Escort,
with high standard of equipment.
Good, predictable handling, light steering and excellent brakes.
Very economical on fuel and fun to drive**

Even when being cornered hard, the Escort 1600 Sport does not roll much. The sports wheels and side strips are standard, as is the black "brightwork".

TO SEE Roger Clark in full cry, doing his balancing trick between 240 bhp and a rutted, muddy forest track is perhaps rather a long way from Mrs Jones setting off to the shops in her Popular. Yet essentially these two Escorts are the same car, and Ford have been very successful at being able to use the first to promote the sales of the rest of the range without in any way making them seem temperamental, unmanageable cars.

The Escort Sport could perhaps be called a "stepping stone" car, midway between the more ordinary 1100 and 1300s, and the much more rare — and expensive — Mexico and RS2000 range. Nevertheless, to the driver who wants the performance and a car which looks the part, it is an extremely good package. There are two versions of the Sport: the 1,297 c.c., 70 bhp model; and the 1,598 c.c. versions. The 1600 Sport and the 1600 Ghia are the last two cars in the Ford range to use the larger of the side camshaft, ohv, Kent engines. The Mexico and the RS2000 use versions of the single ohc engine from the Cortina/Capri range.

Available in two-door form only, the Sport has many extras like low-profile tyres, uprated suspension, halogen head and driving lamps, black external trim, and full instrumentation. To cap it all there is a black coachline, ending in the words "1600 Sport" on the rear wings. In all, a neat and eye-catching car.

As fitted to the Sport, the 1.6-litre engine, with bore and stroke dimensions of 81 x 78mm, develops 84 bhp (DIN) at 5,500 rpm, and 92lb. ft. torque at 3,500 rpm. A Weber twin-choke carburettor is used, along with a special exhaust manifold. Transmission is by means of a standard Escort gearbox, although the ratios are closer to take full advantage of the increased power. On the suspension side, there is the usual Escort package: MacPherson struts at the front, with an anti-roll bar, and a live axle at the rear, located and suspended by semi-elliptic springs. The spring rates are uprated over those of the smaller-engined cars to give better handling at the higher speeds possible.

Performance

Although the rev counter is not red-lined, the handbook is very definite about the rev limits for the 1600 engine. The absolute maximum is 6,500, while 6,300 can be used for continuous operation, should the driver wish to treat the engine in such a manner. Despite the rear axle having comparatively unsophisticated location by current standards, there was no trace of axle tramp, even on the most vicious take-offs, nor could wheel-spin be induced, apart from a faint squeak as the clutch engaged. So, with the fairly high first gear, a careful balance had to be held to prevent the engine revs from falling away for a fast get-away.

We used the maximum permitted revs at first, but found that by

The quarter bumpers and halogen driving lamps are standard equipment, with the sidelamps incorporated in the headlamp units. The vestigial spoiler is painted matt black on the Sport

dropping 500 rpm, around 6,000 for the change points, the best times were recorded. By doing this, 30 mph could be reached without a gearchange and, by just pushing the needle over the 6,100 rpm mark, 80 could be reached in third. These change points compare with maxima in the indirect gears at 6,500 rpm of 33, 61, and 85 mph.

The Sport is no mean contender in the 1600 class, reaching 40 mph in 5.9sec, 60 in 12.3, and passing 80 in 24sec. Yet, despite its high gearing, pulling 18.4 mph/1,000 rpm in top, the car would ease away cleanly from as low as 10 mph in that gear without any fluffing or snatching. On a rather windy day, the Sport lapped the MIRA high-speed circuit at a mean of 97 mph, with a best, down-wind leg of 104 mph, speeds which neatly bracket the 5,500 rpm peak power line, indicating that the gearing is exactly right for the model.

If one has become used to driving less powerful, lower-geared versions of the Escort, there might be an initial impression that the Sport is slightly sluggish. That is until you look at the rev counter and see that

the needle is still way down the scale. By staying in the lower gears for longer, the car suddenly takes on a completely different character and becomes a delight to drive briskly.

Economy

One might well expect a car with this sort of performance to be on the thirsty side, so it came as a something of a surprise that it proved extremely difficult to push the figure below the 30 mpg mark, even on a long, fast run to Yorkshire and back. The overall figure of 34.3 mpg is extremely good, and we found that, with normal, careful use, it was easy to put this figure up to within a few points of 40 mpg.

The constant speed figures bear out the Sport engine's economy, with the car recording 38.4 mpg at a constant 60 and still remaining on the right side of 30 mpg at 70 mph. The tank holds only nine gallons, rather small these days, but with the car's good consumption, this gives a range approaching 300 miles. As we have now come to

A massive air cleaner dominates the four-cylinder, 1.6-litre engine, but this does not prevent the dipstick from being reached. The oil filler cap is slightly awkward to reach

expect with Ford engines, the oil consumption was negligible, and the underbonnet area remained very clean.

Ride and handling

With its big, 175/70-13in. Michelin ZX tyres and uprated suspension, the ride on the Sport 1600 is on the firm side — one might be tempted to say sportingly firm. On smooth surfaces this does not make itself too evident, but it takes only a slightly imperfect road to show up the firmness; the car's shortish wheelbase of 94.7in. tends to accentuate any tendency to pitching. Yet this firmness does not get any worse on really rough surfaces, and the driver feels that the car is under his control all the time. A good point is that the seat suspension rates have been carefully chosen to kill any sympathetic bounce.

For a conventional, front-engined car, where 54 per cent of the weight is over the front wheels, the Escort Sport has delightful handling characteristics. The steering, with 3.5 turns from lock to lock, is light except when parking, yet it remains very precise all the time, with the driver being fed just the right amount of information as to what the wheels are doing and where they are pointing. To all intents, the handling characteristics are neutral, with just a hint of understeer at normal speeds which builds up slowly and progressively as the car goes faster. This gives the Escort a very slight "nervous" feel, and it takes only slight wheel movements for minor changes of direction. But the car is never twitchy and, once the driver has learned that the car is so well balanced, it can be driven very quickly indeed with a minimum of fuss.

The one place that the Escort does tend not to match its normally good handling is on fast, badly surfaced corners, where the back wheels can be set pattering if the car is being driven round hard under power. This gives the effect of a breakaway oversteer, which can be quickly caught and held. Roadholding in the dry is superb, and the Michelin ZX tyres can be made to squeal only under really harsh cornering. In the wet, the lightness of the controls tends to be accentuated, and the back wheels are the first to let go, smoothly on a normal road.

It is in gusty crosswinds that the Sport shows up at its worst, being barged and blown around too noticeably. The quick, responsive steering allows the course to be held but, nevertheless, the car does not feel particularly happy in these conditions

Brakes

Ford use 9.6in. disc brakes at the front, and 9in. drums at the rear, with dual circuits and a vacuum servo on the Sport. The progression on the test was near-perfect, with 20lb pedal pressure giving in-town

Ford Escort 1600 Sport

0.3g check braking, building up to a 1.1g stop with 75lb pressure. It was only when braking fairly lightly from high speed that a slight vibration could be felt, possibly because the disc run out on the test car was towards its upper limit. On the accelerated fade tests the Sport showed up very well, with pressures rising only marginally, with no increase in pedal travel and, for once, no trace of smoke or excessive overworking. Taken overall, they are brakes designed to do the job, without fuss or bother.

The handbrake, a pull-up lever between the seats, held the car easily facing up and down the 1-in-3 test hill. The car moved off quite easily from there, although a careful balance between clutch and accelerator was needed.

Fittings and fixtures

Our test car was finished in a rather lurid green, with matching Cadiz green and black striped cloth seats — not exactly the colour scheme we would choose! Personal preferences apart, the front seats are comfortable, with a sensible amount of location for the spine and thighs. Several people commented that they felt a bit on the hard side, but this proved to be a

good thing on long runs, when the occupants could get out without stiffness or aches. There is ample fore and aft movement for six-footers, and the backrests recline by means of a large, notched handwheel, which allows very small adjustment to be made.

The matching rear seats provide adequate room for adults, especially in terms of headroom. Access is fairly easy, with the front seatbacks being tipped forward after the catches on the outboard edges have been lifted.

The sports steering wheel, trimmed in leather-grained plastic and with matt black spokes, matches the all-black facia well. The driver is faced by three dials, with the rev counter on the left and the speedometer on the right. On the test car, the speedometer was slightly pessimistic, and the mileage recorder under-read by one per cent. The trip odometer is reset simply by pressing the button on the dial face. In the centre is the combined dial for fuel tank contents and coolant temperature.

There are three steering column levers, that on the left being the usual combination for indicators, headlamp dip and flash, and horn. The shorter of the right-hand levers is for the driving lamps, while the

longer controls the wipers, which have two continuous speeds, plus an intermittent action. The screen washers have an electric pump.

Flanking the ashtray are the switches for the hazard warning system and heated rear window, both with their own built-in warning lamps. Ford continue to use a quick-responding air-blending heater but, as on the Cortina Mk IV recently tested, we found that there is too little movement of the temperature lever between cold and warm, so that accurate adjustment is difficult. The dual-level system blows cold air through the demister vents unless the direction lever is set to the screen position, and this ensures that the occupants are never enveloped in a fug of hot air, which can lead to drowsiness. There are the usual adjustable eyeball vents on the facia ends. At town speeds, the booster fan is needed on at its lower speed, but even then it is noisy.

While the facia layout gives a clean and neat appearance, there is a distinct lack of stowage space for items wanted on voyage. There are deep parcel shelves in front of both passenger's and driver's knees — the latter so deep that it is almost impossible to reach the back of it even with the seat belts at full

Two main instruments are speedometer (right) with total and trip recorders, and rev counter (left). Between are the fuel tank and coolant gauges. Column switches are for headlamp dip/flash, indicators, and horn on left. Shorter lever on right is lamps master switch, with wash/wipe switch on longer stem. Only other facia switches are for hazard warning and (hidden) heated rear window flanking ashtray above the radio. The push-button radio is an extra

Specification

ENGINE	
Cylinders	Front; front drive
	4
Main bearings	5
Cooling	Water
Fan	Fixed
Bore, mm (in.)	81 (3.19)
Stroke, mm (in.)	78 (3.07)
Capacity, c.c. (in³)	1,598 (97.4)
Valve gear	Ohv
Camshaft drive	Chain
Compression ratio	9.0-to-1
Octane rating	98 RM
Carburettor	Weber twin-choke
Max power	84 bhp (DIN) at 5,500 rpm
Max torque	92 lb.ft. at 3,500 rpm

TRANSMISSION

Type — Four speed, all synchro

Gear	Ratio	mph/1000rpm
Top	1.00	18.4
3rd	1.43	13.0
2nd	1.99	9.3
1st	3.34	5.1

Final drive gear — Hypoid bevel
Ratio — 3.54 to 1

SUSPENSION		
Front — location	MacPherson struts
—springs	Coil
—dampers	Telescopic
—anti-roll bar	Yes
Rear — location	Live axle
—springs	Semi-elliptic
—dampers	Telescopic
—anti-roll bar	No

STEERING	
Type	Rack and pinion
Power assistance	No
Wheel diameter	14.7 in.

BRAKES	
Front	9.6 in. dia. disc
Rear	9.0 in. dia. drum
Servo	Vacuum

WHEELS	
Type	Deep drawn steel, 4 studs
Rim width	5.0 in.
Tyres — make	Michelin ZX
— type	Radial
— size	175/70SR-13in.

EQUIPMENT	
Battery	12 volt 38Ah
Alternator	28 amp
Headlamps	Halogen, 4-lamp, 230/110 watt
Reversing lamp	Standard
Hazard warning	Standard
Electric fuses	10
Screen wipers	2-speed, plus intermittent.
Screen washer	Electric
Interior heater	Air blending
Interior trim	Fabric seats, pvc headlining
Floor covering	Carpet
Jack	Screw pillar
Jacking points	One each side
Windscreen	Laminated
Underbody protection	PVC wax

MAINTENANCE	
Fuel tank	9.0 Imp gall (41 litres)
Cooling system	9.5 pints (inc heater)
Engine sump	5.7 pints SAE 20W/50
Gearbox	1.58 pints SAE 80EP
Final drive	1.75 pints SAE 90
Grease	No points
Valve clearance	Inlet 0.01-022in. (cold) Exhaust 0.01-0.22in. (cold)
Contact breaker	0.025in. gap
Ignition timing	6 deg BTDC (static)
Spark plug—type	Motorcraft AGR22
—gap	0.030 in.
Tyre pressures	F22; R24psi (normal driving)
Max payload	937lb (426 kg)

Maximum Speeds

Gear	mph	kph	rpm
Top (mean)	97	156	5,270
(best)	104	167	5,650
3rd	85	137	6,500
2nd	61	98	6,500
1st	33	53	6,500

Acceleration

True mph	Time (sec)	Speedo mph
30	3.7	30
40	5.9	39
50	8.6	48
60	12.3	56
70	16.7	66
80	24.0	78
90	38.4	88

Standing ¼-mile: 18.6 sec, 72 mph
kilometre: 35.0 sec, 88 mph

mph	Top	3rd	2nd
10-30	12.1	4.8	5.1
20-40	11.3	7.1	4.5
30-50	10.4	9.6	5.0
40-60	10.2	7.0	6.3
50-70	11.8	8.6	—
60-80	14.4	13.5	—
70-90	23.2	—	—

Consumption

Fuel
Overall mpg: 34.3
(8.2 litres / 100km)
Calculated (DIN) mpg: 27.9
(10.1 litres / 100km)

Constant speed:

mph	mpg
30	55.7
40	50.4
50	44.0
60	38.4
70	30.7
80	25.0
90	21.2

Autocar formula
Hard driving, difficult conditions 31.2 mpg
Average driving, average conditions 37.7 mpg
Gentle driving, easy conditions 41.2 mpg

Grade of fuel: Premium, 4-star (98 RM)
Mileage recorder: 1 per cent under reading

Oil
Consumption (SAE 20W/50) negligible

Brakes

Fade (from 70 mph in neutral)
Pedal load for 0.5g stops (lb)

	start/end		start/end
1	35/40	6	37/42
2	35/37	7	40/45
3	35/40	8	40/45
4	40/40	9	40/45
5	37/42	10	40/45

Response (from 30mph in neutral)

Load	g	Distance (ft)
20lb	0.30	100.0
40lb	0.65	46.0
60lb	0.90	33.4
75lb	1.10	27.4
Handbrake	0.35	86.0

Max gradient: 1 in 3

Clutch
Pedal 33lb and 5.5in

Test Conditions
Wind: 10-15 mph
Temperature: 6 deg C (43 deg F)
Barometer: 29.65 in. Hg
Humidity: 82 per cent
Surface: dry asphalt and concrete
Test distance: 1,036 miles

Figures taken at 9,000 miles by our own staff at the Motor Industry Research Association proving ground at Nuneaton.

All Autocar test results are subject to world copyright and may not be reproduced in whole or part without the Editor's written permission

Regular Service

Change	6,000	18,000	36,000
Engine oil	Yes	Yes	Yes
Oil filter	Yes	Yes	Yes
Gearbox oil	—	—	—
Spark plugs	—	Yes	Yes
Air cleaner	—	Yes	Yes
C/breaker	—	Yes	Yes

Total cost £21.00 £28.03 £33.53
(Assuming labour at £5.50/hour)

Parts Cost
(including VAT)

Brake pads (2 wheels) — front	£10.13
Brake shoes (2 wheels) — rear	£12.28
Silencer	£24.21
Tyre — each (typical advertised)	£21.00
Windscreen (laminated)	£24.21
Headlamp unit	£10.02
Front wing	£16.12
Rear bumper	£13.83

Warranty period
12 months / unlimited mileage

Weight
Kerb, 17.1cwt / 1,914lb / 870kg
(Distribution F/R, 54/46)
As tested, 2,359lb / 1,072kg

Boot capacity: 10.3 cu. ft.

Turning circles:
Between kerbs L. 30ft 0in., R. 30ft 4in.
Between walls L. 31ft 1in., R. 30ft 9in.
Turns, lock to lock: 3.5

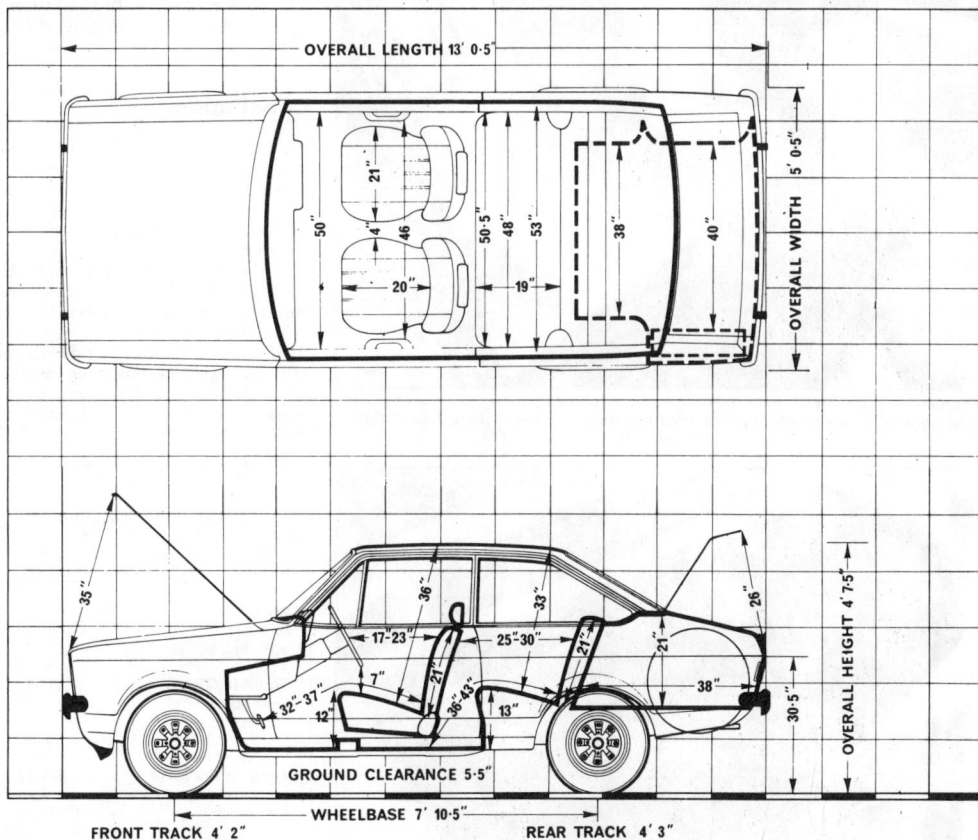

Test Scorecard
(Average of scoring by Autocar Road Test team)

Ratings: 6 Excellent
5 Good
4 Above average
3 Below average
2 Poor
1 Bad

PERFORMANCE	4.33
STEERING AND HANDLING	4.55
BRAKES	4.80
COMFORT IN FRONT	3.83
COMFORT IN BACK	3.57
DRIVERS AIDS (instruments, lights, wipers, visibility etc.)	4.38
CONTROLS	4.13
NOISE	3.67
STOWAGE	3.33
ROUTINE SERVICE (under-bonnet access: dipstick etc.)	3.60
EASE OF DRIVING	4.45
OVERALL RATING	4.09

Comparisons

Car	Price (£)	Max mph	0-60 (sec)	Overall mpg	Capacity (c.c.)	Power (bhp)	Wheelbase (in.)	Length (in.)	Width (in.)	Kerb weight (lb)	Fuel (gal)	Tyre size
Ford Escort 1600 Sport	2,572	97	12.3	34.3	1,598	84	94.7	156.6	62.8	1,914	9.0	175/70-13
Alfetta 1.6	3,519	103	11.5	24.6	1,570	108	100.0	171.0	66.0	2,425	10.7	165-14
BMW 316	3,929	100	12.9	23.2	1,563	90	101.0	171.0	63.0	2,285	11.4	165-13
Datsun 180B	2,600	103	12.5	24.9	1,770	105*	98.4	165.9	63.0	2,248	12.0	165-13
Vauxhall Cavalier 1600GL	2,905	98	14.8	27.2	1,584	75	98.0	174.0	66.0	2,286	11.0	165-13

Gross

AutoTEST
Ford Escort 1600 Sport

stretch. But nowhere is there a place for a packet of cigarettes or sweets, lighter, or sun glasses where they are within ready reach. On the credit side, the ashtray, located centrally in the facia, is big enough for sweet papers *and* cigarette ends.

Living with the Escort 1600 Sport

It is unusual to find an automatic choke on a car with such sporting pretentions, but it does work well. Starting from cold was clean and one was not really aware of an over-fast tick-over or, as the engine warmed up, the effects of the rich mixture cutting out. Hot starting was always at the first turn of the key.

Ford are obviously aware that a driver should be able to see out of the car properly, and the Escort gives excellent all-round visibility. You cannot quite see the rear edge of the boot lid, but the fall-off at the front of the bonnet gives a good idea of the forward length of the car. On the Sport, there are mirrors on both doors, in addition to the dipping interior mirror. The test car was fitted with the remote-control type of mirror on the driving door — £19.03 extra.

Although during normal driving engine revs are kept down by the high overall gearing, the Sport is not noticeably quieter than its rivals. A good deal of noise is transmitted through the suspension, and this becomes particularly loud on coarser surfaces. The door mirrors also seemed to whistle.

With the addition of the auxiliary, halogen headlamps in addition to the standard 7in. halogen headlamps, the Escort Sport has, in effect, a four-headlamp system. On dip beam, the cut off has the typical asymmetric shape while, on main beam, there is an excellent combination of spread and range for fast night driving. The auxiliary lamps cannot be switched separately, and there is no provision for adjusting the level of instrument lighting, which is adequate.

Externally, most of the traditional "brightwork" is finished in matt black, with just the radiator grille surround and cross bar in chrome. The parts getting the most abrasive wear — like the door handles — are electro-chemically plated, while the bumpers, grille, and drip rails are treated with a heat-set black epoxy material. The Sport has only quarter front bumpers, with overriders, instead of the traditional, full-width bumper.

The boot on the Sport is quite large enough to take luggage for four people, and the sides are protected on the left by the spare wheel and on the right by the fuel tank. The minimal wheel changing gear — jack and wheel brace — are stowed in a canvas bag strapped to the side of the spare. The one key is needed to open the boot every time; front doors can be slam-locked.

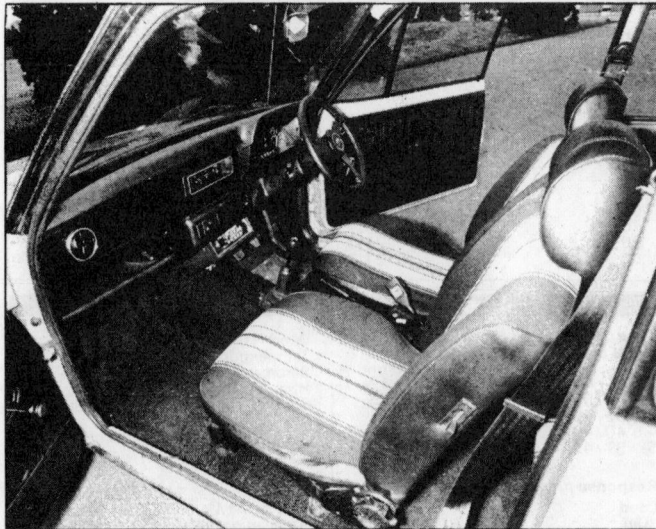
The seats, trimmed in striped fabric, offer good support, with a wheel rake adjustment. The catch allows the front seats to be tilted forward for access to the rear

There is adequate leg room in the rear for adult passengers, with recesses in the car sides for elbow room. The rear shelf is sloped, to prevent items from being placed on it

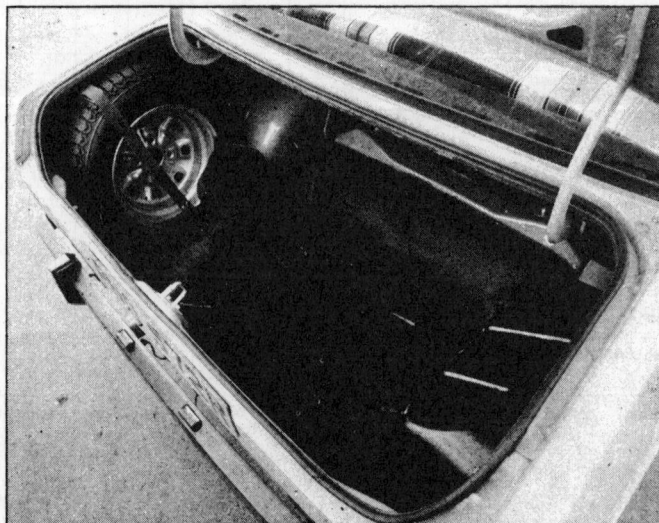
The boot sides are protected by the spare wheel and fuel tank. A mat covers the floor, and the tool kit consists of just wheel changing equipment. Opening is by key only

Under the bonnet, the lid of which has to be propped open, the engine is almost hidden under the huge air cleaner and its thermostatic air intake system. This big air cleaner tends to get in the way of the oil filler cap, but the dipstick is easy enough to get at. Other items needing routine checking, such as the brake fluid reservoir and battery, are easily reached. A block of seven fuses in a panel on the right-hand wheel arch protect most electrical circuits, and there are line fuses under the facia for the heated rear window and the radio; in addition, the Sport has a separate fuse for the auxiliary driving lamps.

Where it fits in

At £2,572, the 1600 Sport is towards the top of the Escort price range, with the 1300GL Estate (£2,614) and the four-door Ghia 1600 (£2,964) the only standard models costing more. These prices include automatic seat belts. The Sport's immediate superior, the Mexico, which uses the ohc 1600 engine from the Cortina/Capri range, developing 95 bhp (DIN) at 5,750 rpm, costs £406 more, while the RS2000, with the 110 bhp version of the 2-litre engine, is £947 more.

Conclusion

To a driver wanting performance and economy, the Escort Sport 1600 makes a good deal of sense. It does take a little time to take full advantage of the car's handling, which has very high limits. The ride is on the sporting side, but the standard of equipment and general comfort do make it a pleasant car to drive, either in town or on the open road. And, by today's standards, the price is highly competitive, especially when compared to some of the more sporting (in appearance alone) foreign rivals. □

MANUFACTURER:
Ford Motor Co Ltd
Brentwood, Essex, CM13 3BW

Prices	
Basic	£2,198.39
Special Car Tax	£183.20
VAT	£190.53
Total (in GB)	**£2,572.12**
Seat belts	Standard
Licence	£40.00
Delivery charge (London)	£46.00
Number plates	£6.50
Total on the Road	**£2,664.62**
(exc insurance)	
Insurance	Group 5

EXTRAS (inc VAT)	
*Laminated screen	£37.26
*Radio	£73.70
*Remote-control door mirror	£19.03
*Fitted to test car	

TOTAL AS TESTED ON THE ROAD	**£2,794.61**

RALLY ESCORT

Ford's Group 4 rally Escort RS1800

In the past decade Ford has established a record of success on international race tracks and rally routes unequalled by any other manufacturer. Latest in a long line of competition vehicles to be developed by Ford is the Group 4 rally Escort RS1800. Basically the same as the regular production Escort RS1800, the Group 4 rally car incorporates modifications to cope with the demands of an international rally.

Underbody protection is increased by the addition of a lightweight alloy shield which covers the underside of the engine and gearbox. The twin overhead camshaft 1800 cc, 16-valve, 4-cylinder engine has its capacity increased to 2-litres and with the addition of two twin-choke, side-draught Weber carburettors — replacing the single twin choke, downdraught Weber — plus modified camshafts and larger valves engine power output is increased to 245 bhp at 7,500 rpm. Additionally, the engine's lubrication system is changed to the dry sump type with the oil reservoir located in the boot of the Escort. The four-speed gearbox is changed to one with five forward speeds and a limited slip differential is incorporated in the rear axle. In order to produce high rates of acceleration with a comparatively low maximum speed the ratio of the final drive unit is also changed.

Many different types of Dunlop tyres ranging from extra wide, hand-cut racing tyres to narrow snow tyres are available to the drivers in the Ford 'works' teams. The light weight alloy, extra wide wheels are enclosed within special wheelarch extensions and both the front and rear suspension systems incorporate gas/oil filled shock absorber units. Rear axle location is also improved and modified by the addition of extra suspension links. In order to provide maximum protection for the occupants of the rally Escort RS1800 a special steel safety cage is built into the car. Full aircraft type seat harnesses are provided for both driver and navigator, a special fireproof bulkhead is built across the rear of the car between the boot and the passenger compartment and the long range fuel tank which is mounted in the boot is filled with plastic foam to prevent fuel spillage in the event of the tank being ruptured. All fuel and brake lines are routed through the inside of the car for protection while the front and rear brake systems are on separate circuits. The ignition can be switched off from either inside or outside the car and a special fire extinguishing system is fitted. In case of fire the system automatically releases a special gas which extinguishes the flames but is not dangerous to the occupants of the car.

The normal production Escort RS1800 is equipped with halogen headlamps and, as a major portion of an international rally is run in darkness, four additional halogen spotlamps are fitted to each of the works rally cars. International rallies form a particularly important part of manufacturers' engineering development programmes. The many improvements which have been brought about in suspension, brakes, lights, comfort and durability in today's production car have been developed in competition. Additionally, the specialist components which are fitted to Ford 'works' rally cars are available to the private owner who wishes to compete in competition. Ford performance parts are marketed throughout Europe through a network of Ford Rallye Sport Dealers.

ESCORT!!

Two Litre Black Top Eater

IF Uniroyal is ever running short of rubber they might like to go out to our test strip where they'll find heaps of the black stuff in a series of neat black lines. This dark mould was deposited by what is going to be one of the excitement machines for the remainder of the decade.

It's the 2 litre Escort dressed up in flashes and stripes looking like a speeding ticket going somewhere to be written. It goes like sand off a shovel and will give most of the wheezing survivors of the Supercar era (and a number of Euro-sporties) a decent run for their money.

If you promise not to say anything to the insurance companies I'll tell you that it's got all the makings of being a latter-day Cooper S.

The plan is to get more younger people buying Escort. Ford has been more than a little concerned that the average age of people buying Escort is 31 compared with, say, Lancer with an average age of 27 years.

Heaven forbid that Escort should become known as an old fogies car — a car for single women to pedal between their unit and the bowling club, a car for men to retire into when inflation has sucked heavily at the nest egg. Mind

you it fills these roles superbly, (in other versions), it's just that Ford would not like to think that Escort was going the blinkered route of the Nomad or Morris Minor.

Fact is you used to be able to buy a rally pack version of the Escort even in the days when the engine was a fairly timid 1.3 litre job. In those days — and even later when Escort went out to 1.6 litres — it was largely a case of sitting behind the stripes, sports wheels and driving lights and dreaming of performance that might have been.

Today there is no time for dreaming — it all happens too fast.

Changes to the 77 Escort are relatively few in number but the result is dramatic. The major task was to shoehorn the two litre Cortina engine into the engine bay. No real headache because it had already been done in Germany and Britain for the RS 2000. So now there's the 1.6 litre engine and 2 litre engine — the latter an option on GL and vans but standard in Ghia.

From the outside it looks like nothing's changed but it was necessary the change the shape of the sump to clear the steering rack and suspension cross member. The radiator is a pace forward with associated underskin

changes. The exhaust is bigger and so is the clutch, while a taller rear axle ratio is used on the two litre version. (from 3.77:1 to 3.54:1). The diff is beefed up and the front suspension struts.

The other major change comes from relocation of the fuel tank. Ford rightly decided that the extra thirst from the two litre would put such demands on the little eight gallon tank that the two litre cars would be calling into service stations with the frequency of a guy with a Japanese bladder.

So they've stamped out their own 12 gallon unit and put it in the floor of the boot. Apart from adding something like 180 to 200 km to the range of the car there is a bonus increase in boot space and an Escort boot is now bigger than that of a Torana. Ah yes, it gets very interesting — didn't think the Sunbird would face a challenge from Cortina and Escort, did you?

The fuel filler is now behind the number plate a la Holden and because that left a hole where the old filler was there is now a blank bolted in which is a reasonable solution as long as the local service station pump jockey doesn't try to get it open with a screwdriver.

Below: Escort rally pack has real "horn car" image. Blacked out bumpers, stripes and real spoiler really look the part. Note the blank over the original fuel filler hole.

Other changes are the use of perforated vinyl seat trim, hazard flasher now standard on all Escorts, Falcon protection moulding standard — although the Ghia's deep section moulding continues.

Pause action wipers are added to Ghia and come with the rally pack and Ghia gets a vanity mirror behind the visor. The speed rating on tyres has been increased (from YR to ZR).

As of August production, air conditioning will be added to the Escort's options and an Escort Ghia with air would have to be the car for today.

The air conditioned car loses some boot space because the compressor (the swash plate kind which does not rob a lot of power) takes up the space normally occupied by the battery. The power pack is therefore re-located in the boot (under a plastic cover) where the old fuel tank was.

Cars fitted with air will get an extra panel under the existing dash containing two air outlets, switchgear (hazard.

driving lights, rear demist etc.) and the digital clock from the LTD. The odds and ends tray houses the ash tray and air conditioning controls.

Cars with air get tinted side glass (tinted back window is standard on all Escorts) and the Ghia gets tinted glass all round anyway).

Actually it's a bit of a tight squeeze

under the bonnet with the big clutched multi-blade fan and bigger radiator necessary as part of the air package.

Which brings us back to the test car and its rally options pack.

Last year we saw 25 RS2000s go on the market — fully imported from Britain by Henry. These had the deformable plastic scoop nose, four

headlights and really looked the part.

Now that Escort has new-found muscle for towing, Mr. Ford is offering a tow pack of bar and gooseneck rated for 1200 lb plus electrical fittings. Make a good little tow car for campers and boats.

Actually the car we drove at the time was not all that impressive on performance but we later concluded there must have been something wrong with it because other testers in other cars romped down their test strips at least a second faster for the quarter than our RS did.

At the time Ford was privately making it public that they would make the RS2000 here and the 2 litre Escort with rally pack is the fulfillment of those whispers.

And it's all that the RS 2000 was except for local differences in fascia, seats (now full foam, by the way) and local sports steering wheel. Volante alloy wheels are available but were not fitted to our test car. It gets a deformable plastic bib spoiler and, in fact, for all intents and purposes it is an RS 2000 with an RS 1800 front on it.

The car was fitted with optional stiffer suspension, an enlarged anti-roll bar at the front and an anti-roll bar at the back plus optional low profile 70 section tyres. The sum total of Ford's handiwork is a car with handling that can stand proud in the company of most European go-machines. It's nimble, agile and fun, fun, fun because there's plenty of power on tap to explore completely the suspension settings of Ford's engineers.

Performance is really quick without being outrageous enough to have the speed bashers thumping their road toll

Below: Boy racer type rubber rear spoiler, together with 2 litre stripe and low slung fuel tank.

Above: Handling of Escort rally pack is superb. As can be seen in the photo, the car has a tendency to oversteer.

statistics from their informed positions at the helms of Austin 1800s.

In fact it all makes sense when put against the *big Mutha* Super Roos with their penchant for slurping hydrocarbons (remember 30 gallon fuel tanks?) laying swathes of very wide rubber and annual insurance premiums to match the lofty annual repayments on the car itself.

For a start this little pony car does not cost the world and does not attract great big registration fees (it's not all that heavy and power is modest by big-banger standards — even though performance comes close to the fire

breathers) Hammering it, we did not get less than 24 miles to the gallon and got a pretty consistent 28 mpg. You'd get 30 mpg quite easily by keeping the second choke on the carby closed.

It pulls well from low speeds in spite of the tallish gearing, has very responsive steering, is easy to see out and so makes an ideal little city car in spite of its big boy racer stipes.

In the country it's got long legs and while the engine works reasonably hard hovering around 140 km/h (at which it cruises well) the sound proofing is good enough to keep you sane on a long run.

It's roomy for its size, has a great big boot, is put together very well, is quiet and comfortable, goes like a train, handles like an Italian love affair and costs less than six grand. Can't think of many cars to match that.

Profile overleaf.

TEST DATA

MANUFACTURER ....Ford Australia
MODEL .... Escort 2 litre GL Rally Pack.
FROM: .... Ford Australia, Campbellfield, Vic.
PRICE (including sales tax) $5152
PRICE AS TESTED $5311
(Includes 70 Series tyres $40, Sports suspension $30, tinted, laminated screen $89.

ENGINE:
Location ....Front
Construction ....Iron
No. of cylinders ....Four
Configuration ....In line
Valve gear ....OHV
Carburation ....down draft-2bbl.
Capacity (litres) 2.0
Compression ratio 9.2:1
Bore x stroke (mm) ....90.8 x 76.9
Power, SAE DIN at 5200rpm (kW/bhp) 70/94
Torque, at 3800rpm (Nm/lb. ft.) 148/109

TRANSMISSION:
Type .... four speed manual
Control location ....floor
Drive ....Rear
Ratios:
1st 3.65:1
2nd 1.97:1
3rd 1.37:1
4th 1.00:1
Final drive 3.54:1

PERFORMANCE:
Speedometer error (km/h):
Indicated: 60 80 100 120
Actual: 59 79 100 120
Acceleration (seconds):
Zero to:
60 km/h 4.6
80 km/h 6.8
100 km/h 10.6
120 km/h 15.7
60-100 km/h 6.8
Standing 400 metres:
Elapsed time (seconds) 16.9
Speed (km/h) 128
Maximum speed in gears (km/h):
1st 55
2nd 107
3rd 140-
4th ....NA
Braking 110-0 km/h (metres/ft):
Best 52.6
Worst 56.2
Average 54.4

BODY/CHASSIS:
Construction .... Unitary
Panel material .... Steel
Weight (kg) 925
Dimensions: (mm):
Length 3978
Width 1595
Height.......................1373
Wheelbase 2406
Track F 1270
R 1296

SUSPENSION:
Front ......Independent, McPherson struts
telescopic shockers, anti-roll bar
Rear .... Live axle, leaf springs,
telescopic shockers anti-roll bar.
Brakes:
Type of system .. Power assisted, disc/drum
Front ....discs.
Rear ....drums.
STEERING:
Type ....Rack & Pinion
Ratio 16.5:1
Turns lock to lock 3.5
Turning circle ....8.9 metres.
WHEELS/TYRES:
Wheel type ....Steel Sports wheels.
Diameter/rim width (in). ....13/5.0 ins
Make of tyre ....Uniroyal
Type .... Steel radial
Dimensions ....ZR 70 S13

PETROL:
Tank capacity (litres) 55
Consumption on test (l/100 km/mpg):
Best 10.0/28.2
Worst 11.5/24.6
Average 10.7/26.4

1. The high compression 2 litre engine is a tight fit in the Escort engine bay.
2. There is not much leg-room in the rear compartment, although headroom is good.
3. Driving position is very comfortable, with good all-round vision.
4. The fuel filler is now hidden behind the number plate.
5. The original fuel filler hole is now filled by a blank.
6. Additional warning lights are badly placed to the right of the steering column.
7. Instrumentation is clear, simple and easy to read.

Motor Manual Road Test

TYPE OF VEHICLE:

FOR all intents and purposes is an RS2000 without .the flexi nose. Two-door Escort with two litre engine and Rally Pack option. Front engine from Cortina driving back wheels. Genuine five seater.

ENGINE/PERFORMANCE:

USES high compression 2 litre engine (new to Australia) on manual cars. Excellent power weight ration (70 kW-94hp) for around a tonne. Very quick when stirred through two stage carby. Does better than 17 sec. for quarter mile. Likes being revved out. Nice and flexible as well. Good economy if not pushed.

GEARBOX:

FOUR speed, floor shift, Beaut gearbox. Right amount of feel in clutch (enlarged) with progressive engagement. Gears easy to find. Changes quite smooth and easy.

BRAKES:

DISCS at the front, drums at the back with power boost. Good pedal, progressive response. A great set of anchors in any situation. Dead straight stops in brake tests. Distance very short, too.

STEERING:

RACK and pinion. Light, very direct. Only three and a bit turns full left to full right. Wonderfully responsive, corrections a pushover. Parking effort firm. Tight turning circle.

SUSPENSION/RIDE:

SUSPENSION tied down really tight as part of Rally Pack option. Ride very firm but not uncomfortable. Tends to pitch over short undulations. Corners dead flat.

HANDLING:

SPOT on. What a fantastic little fun machine. On black top it's all accuracy, maintains lines and really agile. On gravel the tail tends to swing out but it responds quickly to correction. Very twitchy under power, though. Has power and control to make it handle any way you want. Just what you need for driving to and competing in car club events.

PROFILE

CONTROLS/INSTRUMENTS:

MOST controls on steering column wands (set for left hand drive, though). Driving lights, hazard and heater rear window are a lean way on fascia. New brake warning light badly placed. Heater controls not well lit. Fan noisey. Seat rake knurled wheel too close to door to get hand on it. Instrumentation clear, simple, easy to read. Get full, temperature, speedo and tachometer.

INTERIOR SIZE:

ROOMY for small car. Good legroom in front, not bad in back. Headroom front and back is reasonable.

COMFORT/CONVENIENCE:

SEATS well shaped for support, but bit more back support would help. Vinyl perforated for hot weather. Good flow from eyeball vents, heater excellent. One of the quietest cars on the road. Carpet right through. Grab handles, parcels shelf and odds 'n ends bin in console.

VISION:

TWO door version offers better view than four door. Not a blind spot to speak of. Easy parking. Two speed wipers (plus pause) give good wide sweep. Heated rear window. Day-night mirror. Left and right outside mirrors, right hand with remote adjust.

BOOT SIZE/CARRYING CAPACITY:

RELOCATED fuel tank gives bigger boot. Now a cubic foot bigger than a Torana! Fairly high boot lip. Holds four of our Samsonite cases. The smallest to the biggest case.

CONSTRUCTION QUALITY:

SHOWS what can be done when only the one type of car comes down the line. Good trim, good paint. Very tight car. No rattles etc. Very impressive.

BEST POINTS:

PERFORMANCE, handling, quiet running, boot space, equipment level, bright lights and finish.

WORST POINTS:

GEAR lever reach, twitchy handling (don't buy rally pack unless you're an enthusiast). Confusing steering wands.

8. Dash layout is simple, neat and works well. Pedal placement is good.
9. Boot is now extremely large thanks to the relocation of the fuel tank. It will take four of our Sampsonite cases.
10. Driving lights are standard on Rally Pack, but Globe mags aren't. Bumperettes and front spoiler really make the car an eye-catcher.

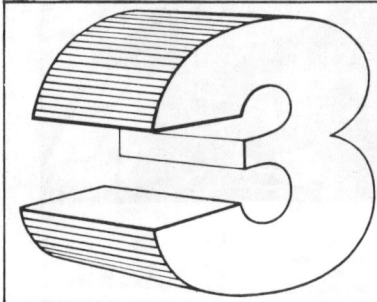

The Escort gets guts

Ford has transformed the traditionally
sluggish Escort into a zappy little performance
car by equipping it with — amongst other
things — a 2-litre engine

"BUT, GRANDMA, what big teeth you have."

"All the better to gobble up the kilometres, my dear."

"And what are all those stripes and spoilers and air dams you're wearing?"

"All the better to get across my new image, my dear."

Quicker than you, or even Roger Clarke, could wander through the woods, Ford has transformed old Grandma Escort into a wolfy little performance machine.

From being the small-car slug it was yesterday, suddenly the Escort now offers about the highest performance-for-the-dollar ratio of any car on the Australian market.

Till now, that rally-bred image of the Escort was something Australians could only read about. With the exception of a handful of early twin-cams and a limited run of RS2000s, we had to content ourselves with 1300cc Escorts perched on bicycle wheels.

But that has all changed with the option of the 2-litre engine. At the same time as giving the Escort proper power, Ford has slotted options and new standard equipment which make it a proper car.

Even enthusiasts will find appeal in the Escort. And that is obviously part of Ford's plan.

In England, the backbone of club motorsport is the Escort, which is available there in a variety of performance versions. Till now, only Japanese cars made much sense here for motorsport amateurs.

It would please Ford greatly if that changed, and there is no reason why it won't.

Not that car club-type people buy lots of cars. Rather, they are recognised as opinion leaders: a lot of buyers question people who are obviously interested in cars about an intending purchase. And that means enthusiasts can influence a very wide section of the market.

Perhaps it is significant that the other weekend we noticed no fewer than three 2-litre Escorts at a car-club motorkhana.

Not that the 2-litre Escort is an instant racer. In its present form, the overhead cam engine develops 70 kW, about 9 kW down on the figure quoted when the RS2000 was first released.

And Ford has chosen to cut back acceleration in favour of better economy at high-cruising speeds by using a relatively high ratio 3.54 to 1 rear axle. This gearing lets the engine loaf along, with 2000 rpm corresponding to a road speed of 60 km/h. But that is still a good enough

power-to-weight ratio to give the car very respectable performance.

The 2-litre engine is flexible enough to let it pull fairly low in top, but naturally the particularly high final drive does mean that a lot of traffic work is done in third gear.

When you use the gearbox to advantage, you do find the overall higher ratios mean that, for a given situation, you are often in a cog lower than you might be in the more conventional buzz-box.

The third is great through moderately fast twisty stuff and for overtaking at highway speeds.

Our rally-packed test car came with a tacho, which — rather interestingly — wears no red-line. However, power drops off fairly rapidly around 6000 rpm so there is no point in exceeding that limit, which is really quite enough for a cooking 2-litre engine.

Competition versions certainly rev higher, but there is a lot to be done under the skin to make that a safe practice.

The gearbox itself is the same old Escort unit, with a change so light you can practically throw the lever from one slot to the next. However, the lever has a poor relationship to the seat and third gear is an uncomfortable reach for most drivers. Ford would do well to make the stick longer and bend it back towards the driver.

Now that it has the power to appeal to people who want performance, the Escort is available with options which give it a "competition" appearance.

These include a glass-fibre air-dam below the front bumper, a spoiler on the boot lid and 5.5in-wide alloy wheels.

Our test car was tricked up in this way and was also fitted with the optional sports suspension pack which comprises a stiffer front anti-roll bar, a rear anti-roll bar and stiffer springs and shock absorbers.

We half-expected the extra weight of the bigger engine up front to spoil the previous "driveability" of the Escort. But, certainly in sports suspension form, it works better than ever. In fact, the handling is the type that sends you looking for corners.

The final limit of roadholding may not be in the ultimate bracket, but you can fling the Escort at corners and know it is going to go exactly where you plan it to go.

Occasionally, a bump will cause the cart springs at the rear to have a bit of a hiccup, but even then a flock of the

The heart of it all is the ex-Cortina 2-litre engine.

wheel brings it back on line.

On dirt, it's magic and even the standard car does possess a lot of the marvellously predictable behaviour that is so evident in rally versions.

On tar, there is slight understeer with the power on, but the cornering line can be tightened merely by lifting the throttle. If you do hang the tail out, correction is a flick of the wrists. It's well balanced enough to be held on opposite lock on a wet road without any sign of drama.

It is easy and fun to drive. The sports suspension pack does make the ride a little harsher, but the benefits are more than worth this slight compromise.

Not so pleasing, on our particular test car, were the brakes. When used hard, they vibrated badly and needed fairly high pedal pressures. They stopped the car satisfactorily, but they did not inspire confidence. In fairness, the test car had probably had a pretty harsh work-out.

To match the up-market move of the Escort, the interior has been revised slightly, but items like the door trim and the seats give away the vehicle's true age.

There is nothing particularly wrong with the function of the interior styling, but it is dated. There is no glovebox, but there is a centre console in addition to a parcel tray below the dashboard.

The instruments are grouped in front of the driver and, on our optioned-up test car, comprised a speedo with re-settable trip meter, tachometer, temperature gauge and fuel gauge.

Four-way hazard flashers and an electrically-heated rear window are standard.

A feature of the $360-odd rally pack (or Rallye Pack, as Ford calls it) is an intermittent windscreen wiper mode with a seven-second delay between wipes.

There is a difference

JUST HOW far removed is the 2-litre Escort from its rally-winning cousins?

Well, that depends on how serious you are about winning rallies. If you are content to have a careful run in club events and are prepared to slow up over the really rough bits, some reasonably mild modifications should keep it in one piece.

The engine, transmission and diff could be left standard. A substantial sump guard, mounted to the frame and not the cross member, would be a must. Now the tank is slung low under the boot, you would need a guard for that, too. Better, it should be relocated.

To use a near-standard car like that, you would have to drive very much within limits. But there is just about no end to what you can, and really must, do if you are serious about rallying an Escort.

The RS2000 Escort is basically the same as the Australian 2-litre, so the modifications staffer Jim Sullivan has made to his NBN TV/Kloster Ford RS2000 give some idea of what is necessary.

The car has an engine built for reliability and turning out around 105 kW on dual downdraught Webers. The engine uses special con-rods, pistons, valves and a

toughened crank for obvious reasons.

The suspension uses special Bilstein struts, stronger coil springs, different rear springs and Bilstein rear shock absorbers. The strut towers have strengthened mounting points.

The major component change is the rear axle, which is a "mini-atlas" unit that is a lot more robust, and is fitted with a special strengthening plate.

It uses a limited-slip differential with a 4.6 to 1 ratio.

Larger rear brake drums from a Capri V6 are used, while ventilated discs replace the normal solid discs at the front.

A special front stabiliser bar mounting kit is used, which has much more substantial mountings.

A full alloy roll cage is mounted to a number of points inside the Escort.

However, the NBN TV/Kloster Ford car is only about a 50 percent job compared to the full-house version of Colin Bond. His beautifully-prepared car has the body extensively re-welded and plated. It also uses an entirely different front cross member.

And it still does not go as far as the full-house works rally cars . . . □

A continuing annoyance with the Escort is the use of a steering-column stalk as a light switch. There is no reason why this third stalk could not be merely a simple dashboard-mounted switch, avoiding the present situation where it is extremely easy to knock the switch to "on" when switching off the ignition. It is the right formula for a flat battery when the car is left parked for the day.

With its new-found performance, the Escort also deserves better seats. They are reasonably comfortable, but are too wide and shapeless, providing little lateral support. The rear seat is comfortable enough, but has precious little leg room.

Maybe in anticipation of enthusiastic driving, rally-style panic handles are mounted above each passenger position.

With power to carry a little more luggage, the latest Escort has a larger boot. This has been achieved by putting the fuel tank — which is also larger — beneath the floor, leaving clear the mudguard space formerly occupied by the tank.

The previous filler hole in the mudguard has been closed off with a simple blanking plate and it is great sport watching service station attendants try to "open" it.

The old Ford bogey of "finish is poor finish" is still evident in the Escort. The interior of the boot on the test car was particularly rough and a number of body rattles were evident.

On the other hand, general noise levels are particularly low. In part, this is due to a much less busy engine.

Particularly impressive was the fuel consumption of the Escort. The nature of the car invites fairly vigorous use and yet it still returned an overall test figure of 9.34 litres per 100 kilometres.

With a little more refinement, the 2-litre Escort would have to be a very good argument for not spending twice as much on a European exotic . . .

EGAD! THEY'VE GIVEN THE
ESCORT 54 PER CENT MORE POWER...

POWER
TO THE PUNY

Suddenly, a Ford Escort is faster than an Alfetta GTV or a Lancia Beta coupe. They've shoe-horned in a 2-litre Cortina engine and turned the 97 lb weakling into a ball of muscle.

IN EIGHT years in Australia the Ford Escort must have set some kind of record for having the greatest number of engines and engine capacities.

By our count the Australian Escorts have had six engines in eight years — a 1.1-litre Kent, a 1.3-litre Kent, a 1.56-litre Lotus-developed twin-cam engine, a 1.6-litre Kent and now the Cortina's two-litre single overhead camshaft engine. Apart from the Twincam Escort, which was a high performance "special" sedan, the gradual growth in engine capacity has reflected Ford's simple desire to maintain a good level of performance in its smallest car in the face of increases in body weight and the fitment of power-sapping anti-emissions equipment. FoMoCo, more than any volume manufacturer we know, hates the thought of one of its products being rather weak-kneed.

Since the Escort II appeared in 1975 there have been large transfusions of power for the Escort. At first the car had only the 1.3-litre Kent engine and we called it underpowered. Then the 1.6-litre Kent came in as an option and we reckoned that car's performance was adequate. Now the 1.6 is the base engine and you can have a two-litre engine for a bit over $200 extra. For that you get a whopping 54 percent more engine power. The two-litre's power is almost double that of the 1.3. The issue now is certainly not one of lack of power. It is now whether Ford hasn't used some overkill to lay to rest the Escort's performance deficiencies.

The new engine, bigger and heavier than the 1.6, has had to be shoe-horned into the Escort. The radiator has been moved forward and surrounding inner panels have needed modification; there has had to be a new sump to squeeze the engine around the Escort's steering gear and front crossmember. There is now a bigger diameter exhaust pipe and a bigger capacity front muffler and the clutch has been increased to the

ROAD TEST

Above: Escort rally pack with two-litre engine corners hard with a little understeer, very little body roll. Across the ground, it's one of the fastest two-litre cars you can buy.

Right: The muscle of the two-litre Escort comes from this SOHC Cortina mill. The Escort radiator had to be moved forward and the shape of the sump modified to fit around steering and structural components.

Centre: Escort two-litres have a bigger boot than the older models because the tank is taken out of the boot and located under the floor. Filler is behind the number plate, not on the right hand side as before. Filler tube and cover (at left of boot catch) intrudes into boot space.

Bottom: Escort's driving position is quite good, though the wheel is quite high and protrudes a long way from the dash. This is the Ghia version which has excellent imported seats. Even the locally-made seats in lower-line models are very comfortable.

Cortina's diameter to cope with the power.

The gearbox has needed no change but the rear axle ratio has been changed from 3.77:1 in the 1.6 to 3.54:1 to give the car longer legs. As well, the differential side gears are now made of "better material"

The front suspension struts are made stronger to cope with the extra engine weight and the front spring rates and anti-roll bar are stiffer. Based on Ford's figures there seems to be no vast weight difference between a 1.6-litre and a two-litre car of equal specification, but the fact that the radiator has been moved forward indicates that there is a greater concentration of weight over the front wheels. Besides, the 54 percent extra power is likely to impose heavier loads on the whole factory.

The two-litre Escort comes in two and four-door GL versions and a four-door Ghia version. Prices start at around $5150 (list) at the time of writing for the GL two-door and stretch up to $6400 for the four-door Ghia auto.

The GL Escort, which is the "Kingswood" version, has reclining bucket seats, carpets, a rear window demister, radial tyres, hazard flashers, quartz halogen headlights, an AM radio, a cigarette lighter and various distinguishing "mouldings" and wheel covers so your neighbors will know you didn't buy cheapest.

The Ghia is a House & Garden version of the Escort with shag-pile carpet, imported seats with fabric facings and comprehensive support for the body, wood on the dashboard fascia, a vinyl roof and all the gadgets and gargoyles that make it a junior LTD. About the only useful LTD features the Ghia lacks are air conditioning and a stereo tape player (and both of those are options).

The two-litre engine comes in two versions, as it now does in the Cortina. There is a high compression (9.2:1) version which produces 70 kW and goes into manuals, and a medium compression version (8.2) which puts out 64 kW and propels all automatics. Originally the two-litre, an imported engine, came only with the lower output but the more powerful mill has been shown to comply with Australian design rules when coupled to a manual trans, so it is used in manual cars.

Although the specification of the Escort is now pretty high, the list of options available on top of a GL and a Ghia is still quite long. Most interesting for the GL is a rally pack which consists of decals, matt black this and that, twin

Above left: The two-litre Ghia's interior is neat with velour seat facings, wooden dash fascia, cut-pile carpets, soundproofing and a glovebox (which other Escort models don't have).

Left: This is the rally pack steering wheel and instrument spread. Control layout and instrument visibility are exemplary, steering wheel is brilliant to look at and use.

driving lights, bumperettes at the front, stiff suspension, full instruments (including tacho), bright exterior colors and a fabulous small diameter, thick-rimmed steering wheel.

You can now buy integrated air conditioning in Escorts, Globe Volante alloy wheels and you can option on the front air dam which is also part of the rally pack kit.

The two-litre engine makes the Escort — either manual or auto — a mighty quick little car. In manual form (and thus with the more powerful engine) the car will pull a 17.5 second standing 400m which is excellent for a clean-engined car. It's about half a second faster than an Alfetta GTV or a Lancia Beta coupe and those cars used to be the standards by which two-litre performance was judged. The Escort is faster than practically everything short of a five-litre V8. Its strength is obviously in the first three gears; it gets to 90 km/h in 9.3 seconds which is only a second slower than a Holden GTS five-litre takes. After that, of course, the difference increases.

The automatic, with its lower compression engine and greater power losses gets across a standing 400m in 18.9 seconds and a 10th or two better if you use manual over-ride to postpone the first to second and second to third gearchange. The auto is quite quick — within a second of the manual's time — in its acceleration to about 80 km/h, but after that it tends to taper off. But it's quick for a small auto, just the same.

As important as the stopwatch figures is the two-litre Escort's feeling of effortlessness. Its gearing is taller than that of previous models and consequently it doesn't thrash and whir as much for a given amount of activity. And that given amount of activity requires considerably less of a push on the loud pedal, too. A cruising speed of 130 km/h is child's play for either car's mechanicals, but wind and road noise build up to a point at about 140 where you wouldn't go any faster unless it really mattered.

Escorts have been predictable and easy-to-drive cars for as long as they have been cars. The two-litre engine in the Escort just makes things easier since it adds substantially to the car's flexibility for a minimal increase in front end weight and consequent steering effort. The steering is much lighter and more accurate than in almost all other small cars we drive regularly (because most of them are Japanese and they still can't build precise steering over there except in Honda factories).

The two cars we drove for this story were a Ghia auto with standard (softer) spring/damper rates and a rally-packed GL manual. The manual car is a firm-riding little firebreather which handles predictably and sticks to the road like the white line. In the dry it corners with a whiff of understeer which builds up as you get more daring until the car is threatening to let go at

the front. It's not really possible to hang the tail on bitumen corners, in spite of the car's ample power, because at eight or nine 10ths of cornering effort the inside rear wheel loses adhesion and the power is wasted in wheelspin.

The rally-pack test car also had the optional Volante alloy wheels and fat tyres which gave great dry weather grip. Unfortunately they weren't as good in the wet. We experienced both front and rear wheel slides at quite low speeds and not all of them predictable. Fortunately the sensitive steering and compact dimensions of the Escort made for easy recovery.

The Ghia was simply a softer version of the rally-pack car. It displayed the same handling characteristics except that its steering responses were a little slower and it rolled more when cornered hard. The standard car has an anti-roll bar at the front while the rally-pack has them at both ends (and the front one is stiffer).

The rally-pack suspension isn't really designed for comfort and consequently it doesn't make a good-riding car. It jigs and jumps quite a lot on city street bumps taken at low speeds but smooths out as speed builds. On the two occasions we carried rear passengers in the test car there were mild complaints about the ride being bumpy. The imported RS2000 Escort (one of a batch of 25) which WHEELS drove 18 months back had a similar hard ride to the home-grown version but it seemed to have stronger damping. It jumped about less and seemed generally to have its springs under better control on rebound. It needs to be borne in mind, however, that the latest test car had been used as a survey car for the '77 Southern Cross rally and while in quite good condition had unquestionably been "used".

The Ghia is simply softer-riding. The sound-proofing, underfelt and carpets mean that it is a very quiet-riding car. It is as susceptible to rear axle wind-up over ripples as any other conventionally suspended car (Gemini, Corolla, Datsun 120Y). But it doesn't have the Gemini's front-end mushiness or the Datsun and Toyota's over-hard, under-damped character, and so it feels a cut above them.

Taken all round, Ford has made a good move by dropping the two-litre SOHC engine into its Escort range. It has given the manual car the power to take full advantage of the firm rally-pack suspension (the Escort is now one of the two or three fastest two-litre cars available here) and it has given the Ghia automatic the kind of effortlessness which is needed in a mini-LTD.

More fundamentally, it has given the Escort a helping of that "Ford-aggressiveness" which the Falcon and Cortina ranges already have — mainly by virtue of their big engines. The Escort is a REAL Ford now.

SPECIFICATIONS ▶

SPECIFICATIONS

MAKE . . . . FORD
MODEL . . . . Escort 2.0-litre Ghia and GL
BODY TYPE . . . . 2 and 4 door sedans
COLOR . . . . Silver grey (Ghia), yellow (GL)
PRICE:
Basic $5151
As tested . . . . Ghia $6391, GL $5825
OPTIONS FITTED: Ghia: none; GL: rally pack ($323), AM-FM radio ($152), Volante alloy wheels ($199).
ENGINE:
Cylinders . . . . Four
Valves . . . . Overhead, single overhead camshaft
Carburettor . . . . Twin-throat, downdraught
Compression ratio . . . . .9.2 (manual) 8.2 (auto) : 1
Bore x stroke . . . . 90.8 x 76.9 mm
Capacity . . . . .1.998 litres (121 cu in)
Power at 5200 rpm . . . . 70 kW, 94 bhp (man)
Power at 5300 rpm . . . . 64 kW, 85 bhp (auto)
Torque at 3800 rpm . . . . .148 Nm, 109 lb-ft (man)
Torque at 3500 rpm . . . . .140 Nm, 103 lb-ft (auto)
TRANSMISSION:
Type . . . . .Four speed all syncro (GL), three speed auto (Ghia)
Gear lever location . . . . .Central, auto has T-bar

RATIOS — MANUAL:

	Gearbox	Overall	km/h per 1000 rpm	mph per 1000 rpm
First	3.65:1	12.92:1	8.01	4.98
Second	1.97:1	6.97:1	14.87	9.24
Third	1.37:1	4.85:1	21.35	13.27
Fourth	1.00:1	3.54:1	29.27	18.19
Final drive	3.54:1			

RATIOS — AUTO:

First	2.47:1	8.74:1	11.86	7.37
Second	1.47:1	5.20:1	19.90	12.37
Third	1.00:1	3.54:1	29.27	18.19
Final drive	3.54:1			

CHASSIS AND RUNNING GEAR:
Construction . . . . All steel, unitary
Suspension, front . . . . MacPherson struts, anti-roll bar
Suspension, rear . . . . Live axle, semi-elliptic springs, anti-roll bar on rally pack
Steering type . . . . Rack and pinion
Turns I to I 3.5
Turning circle . . . . .8.9 m (29.2 ft)
Brakes, type . . . . Disc (front), drums (rear)
DIMENSIONS: (GL)
Wheelbase . . . . .2406 mm (94.7 in)
Track, front . . . . .1270 mm (50 in)
Track, rear . . . . .1296 mm (51 in)
Length . . . . 3978 mm (13 ft 0.6 in)
Width . . . . 1595 mm (5 ft 2.8 in)
Height . . . . 1373 mm (4 ft 6.1 in)
Fuel tank capacity . . . . .54.6 litres (12 gallons)
Kerb mass (weight) . . . . 925 kg (2039 lb)
(Ghia) as for GL except:
Length . . . . 4059 mm (13 ft 3.8 in)
Height . . . . 1372 mm (4 ft 6 in)
Kerb mass (weight) . . . . 1003 kg (2207 lb)
TYRES: (GL)
Size ZR70S13
Make fitted . . . . .Uniroyal 180/70 steel radials
(Ghia)
Size . . . . 175/70 SR 13
Make fitted . . . . Michelin ZX

POWER TO THE PUNY

PERFORMANCE

TEST CONDITIONS:
Weather . . . . Fine
Surface . . . . .Castlereagh Dragstrip
Load . . . . Two persons
Fuel . . . . .Super

SPEEDOMETER ERROR:

Indicated km/h	50	70	90	110	130
Actual km/h GL	50	70	89	108	128
Actual km/h Ghia	49	69	88	107	127

FUEL CONSUMPTION ON TEST:
GL:
Check one . . . . .9.37 km/l (26.4 mpg) over 177 km
Check two . . . . .10.33 km/l (29.1 mpg) over 313 km
Ghia:
Check one . . . . .9.76 km/l (27.5 mpg) over 214 km
Check two . . . . .8.52 km/l (24.1 mpg) over 173 km
MAXIMUM SPEEDS:
Average all runs . . . . (GL) 175 km/h, 109 mph
(Ghia) 160 km/h, 99 mph
IN GEARS: GL
First . . . . .48 km/h (30 mph) (6000 rpm)
Second . . . . .88 km/h (55 mph) (6000 rpm)
Third . . . . .129 km/h (80 mph) (6000 rpm)
Fourth . . . . .175 km/h (109 mph) (6000 rpm)
IN GEARS: Ghia
Drive:
First . . . . .57 km/h (35 mph) (5200 rpm)
Second . . . . .106 km/h (66 mph) (5400 rpm)
Third . . . . .160 km/h (99.4 mph) (5450 rpm)
Held:
First . . . . .71 km/h (44 mph) (6000 rpm)
Second . . . . .119 km/h (74 mph) (6000 rpm)
Third . . . . .160 km/h (99.4 mph) (5450 rpm)

ACCELERATION:

Through the gears:	Man	Auto
0-50 km/h	3.8	4.9
0-70 km/h	6.3	7.4
0-90 km/h	9.3	11.2
0-110 km/h	14.0	16.3
0-130 km/h	20.2	25.2
0-150 km/h	—	—

In the gears:	Manual			Kickdown
	Second	Third	Fourth	(Auto)
30-60 km/h	3.2	4.8	7.5	3.1 secs
40-70 km/h	3.2	4.9	7.4	3.5
50-80 km/h	3.5	5.1	7.5	4.6
60-90 km/h	4.2	5.3	7.8	5.6
70-100 km/h	—	5.6	8.1	6.4
80-110 km/h		6.2	8.8	7.7
90-120 km/h		7.2	9.7	10.1
100-130 km/h		9.2	10.9	12.9
110-140 km/h				

STANDING START (0 — 400m):
Fastest run . . . . .17.5 secs (man) 18.8 (auto)
Average all runs . . . . .17.5 secs (man) 18.9 (auto)

IMPRESSIONS

FORD ESCORT RS2000

"More than enough power to satisfy the potential Harvey Yaps of the world," says P.H. Cheah after trying the top Escort.

Mention the letters RS to anyone who knows something about Ford's range of cars you will probably get a positive reaction. Ford have promoted the RS image so well that many people, even here in Malaysia, think of the prefix as something that spells performance, handling and perhaps even looks. After all, it's the RS Escorts that have won numerous victories in races and rallies the world over and there won't be too many who would argue that the RS Escorts are perhaps the most successful competition saloons in the world.

Such an image wasn't born overnight and to be sure, Ford worked long and hard at it. Their competitions department is just about the most professional, and possibly the best there is and these cars are strong, fast and tough enough to take almost anything a rally or race driver might subject the poor beast to.

In Europe, those lucky people can buy one of three RS models: the 1.6 litre RS Mexico, the RS 1800 and the RS 2000. Production of the complex 16 valve twin-cam RS 1800 has ceased but the other two continue in production. The RS cars are no longer built at Ford's Advanced Vehicle Operations (AVO) establishment in the UK; instead they are built in Germany, alongside their more mundane compatriots.

The subject of this test is the more distinctive RS 2000, a car that Ford of Europe's Harry Calton managed to arrange for tests within only 20 minute's notice.

A Special Escort

Being an Escort, the RS 2000 is a competently styled saloon with a low waist line and a large glass area. The big difference between the RS 2000 and the standard saloon is the frontal styling as the car has a 'droop snoot' plastic nose incorporating four round halogen headlights. The grille itself is all black with the Ford corporate badge in its centre. Turn indicators are placed just under the front bumper, not exactly the best position if minor accident damage is to be avoided.

Black is also the theme of the car's bumpers, window surrounds, door handles and the rear spoiler. At the rear, the car is typically Escort with the simple tail light designs and black again used to finish the tail area. The big addition is the use of the soft, rubberised spoiler.

Overall, the car is pleasing and sufficiently distinctive to let others know that this is a rather more special Escort. The alloy wheels and those fat Pirelli steel

belted tyres help to give that purposeful look to the car as well.

Aside from the modifications to the suspension mounting points and the 2 litre Pinto engine under the bonnet, the car is a mechanically straight forward Escort. The front mounted OHC engine churns out 110 bhp at 5,500 rpm which should endow the car with a lot of performance.

Suspension is independent in front by MacPherson struts and there's a thick anti-roll bar to control body roll. The rear utilises leaf springs on a well located live axle and all suspension mountings have been strengthened to take the additional power and the kind of pounding an RS model is likely to encounter in its life. An anti-roll bar is also used at the rear.

The four-speed gearbox has synchromesh on all forward gears, stearing is by the usual rack and pinion system and servo assisted disc brakes are used up front with self-adjusting drums at the rear. The handbrake is located between the front seats and they operate on the rear wheels.

Comprehensive instrumenta tion but....

Interior comfort of the car depends on whether you sit in front or at the rear. The front seats are super as they provide excellent lateral support and thigh support is good as the cushion stretches further out than normal. The seats are larger than even the Escort Ghia's and in this way, they steal space from the front and the rear.

The driving position is excellent with well positioned pedals and steering. The thick rimmed, padded steering wheel is a pleasure to hold as well. Controls are typical Ford in that the three steering column stalks operate the lights, wiper washers, main beam, flasher horn and indicators. On the fascia are the hazard warning, rear screen heater and rear fog lamp switches as well as the cigar lighter. Ventilation controls are placed in the centre of the fascia with the two speed blower switch placed on the left of the slide controls.

In front of the driver, the familiar Escort instrument binnacle contains the rev counter, the speedometer, the oil pressure, fuel and temperature gauges. Below the smaller dials, there's a row of warning lights consisting of the ignition, main beam, turn signal, oil pressure and precious nothing else. It's a pity that Ford have not incorporated items like warnings for hand brake 'on' or door ajar or at least low fuel level.

The instruments themselves are an object lesson in legibility as they are clearly calibrated with the non-reflec-

hand release knob for the front seat backrest came off a couple of times, but good nonetheless. At least everything fitted well and the doors closed with a most reassuring solid sounding clunk.

Delightful high performance

Put a fairly large engine into a small, relatively light body and you invariably end up with a potent machine. If the car, in standard form already possesses proper reserves in handling and roadholding, the package can be very entertaining indeed. In the case of the RS 2000, entertaining is the most appropriate word to use.

The Escort 1600 Ghia and Sport are no slouches, so just try to imagine how much faster the 2 litre engine can be in the Escort shell. Power is put to the

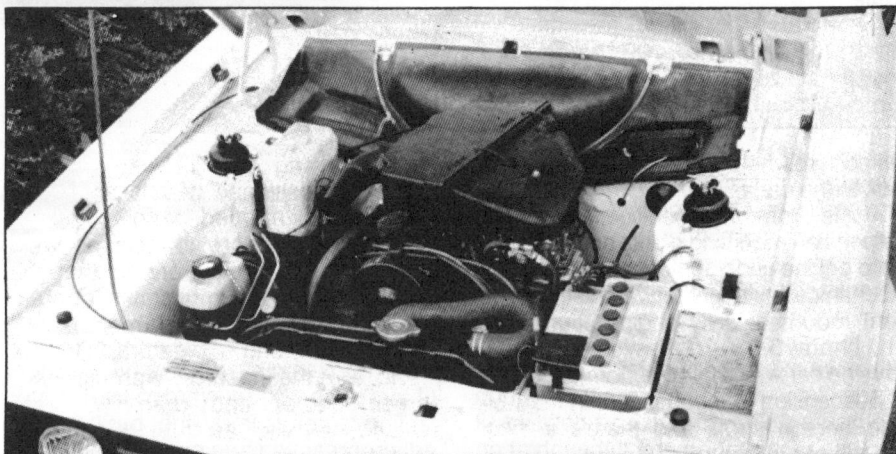

tive angled glass doing the job it was designed to do. In fact the fuel gauge calibration is clear enough to tell you when you've got a quarter tank or three quarter tank of fuel aboard as it is calibrated in the four quarters quite clearly.

With its well placed legible instruments and the ergonomic placement of controls, the RS 2000 cockpit is truly one for the driver's driving pleasure.

Compact interior

The interior is finished in black which I found rather cold and uninviting as not only are the seats and door trim in black, the headlining and the sun visors are similarly finished. Some people may call this sporting, but I don't. The seats are cloth-covered and as mentioned before, the front ones are comfortable.

Those seated at the rear are not so well off. The rear seat is rather shapeless but is fairly comfortable for two. Put three in and the instrusion of the rear wheel arches take valuable space and if you have the front seats right back, there's practically no room for feet: I suppose there's not much else that Ford could have done in the short overall length of the car and its wheelbase. That's one of the main handicaps of conventional drive but then one can argue that the RS 2000 would seldom have anyone in the rear anyway. Trouble is, the same applies to the cheaper Escorts as well.

Still, whatever my thoughts about the all black interior, it has to be said that the quality of fit and finish is good, not outstanding mind you, as the left

rear wheels quickly and efficiently and the car is capable of reaching the 110 km/h (70 mph) limit in less than 10 seconds. The engine pulls very strongly through the gears and even at high cruising speeds on the motorway, any additional acceleration needed for overtaking is provided without having to change down to a lower gear.

Ford claim a top speed of 176 km/h (110 mph) on the RS and I have no reason to doubt their claim as test reports in other magazines have shown top speed figures close or above the claim.

The engine is a very willing performer and will rev up quite freely but it does so with a degree of noise that does intrude into the passenger compartment. While the engine note is not unpleasant, the car does not seem to have the quiet refinement of the Escort Ghia.

Other test reports on the RS 2000 have the zero to 100 km/h (62.5 mph) times at around 8.6 seconds and while I could not take such performance time, I found the car was capable of accelerating with the sort of liveliness that makes overtaking a safe and delightful exercise. On hard acceleration, there was no trace of any axle tramp although the car was wheelspinning like mad when the power was turned on.

The gearbox is good but somehow I felt it wasn't up to the usual Ford standards. While the movements through the gates were smooth and positive, it felt just a little bit notchy although at no time did I experience any baulking from the box. The short throws required to enter each gear do help to make the box easier to use but I can't help feeling that this box has been designed for rougher use than the standard

Escort boxes, as it feels rather stronger (if that's the correct word to use).

The clutch is fairly light but it provided strong action and could cause a jerk when moving off from rest to someone who isn't used to it. Anyone raised on a steady diet of Japanese cars may find the RS 2000 too heavy, in both clutch and gearbox but then he would not be too interested in such a car anyway.

Fuel economy isn't expected to be a strong point of any performance car so it came as a surprise that the test car returned a creditable 27 mpg during the period of the test and this was driving on the open road, the motorway and in the city of London itself, where you tend to use the gearbox a lot.

An enthusiast's car

The Escorts are generally better handling cars than most of their more mundane Nipponese rivals so the RS 2000 should be something more 'special' and you can bet your sweet life it is. The suspension is set for handling, rather than ride and the additional stiffening to the suspension mountings must all contribute to the car's superb handling.

And superb it is; push the car into a corner at high speed and the car responds beautifully, without drama and always obeying the driver's every command. There is so little body roll that passengers may not sometimes realise that the car is being cornered at higher speeds than they are used to.

The positive feel of the steering helps, of course, as you always know what the road surface is like and if you lift off in mid-corner, there is a small degree of tuck in, but the car remains almost totally unperturbed by such an action. The RS 2000 remains almost completely neutral throughout its speed range although with the available power, it can break into oversteer if the car is pushed up to and over its limits. Yet when this happens, it is gradual and corrections made at the wheel result in the sort of response that makes driving the car in country lanes a pleasure, providing of course you do not encounter another vehicle coming from the opposite direction. Those lanes tend to be a little narrow, so one must be a little cautious.

Like any modern car, the Escort has more reserves of roadholding than most would require but the RS 2000 is an enthusiast's car so it must be better than average and it is. Those Pirelli tyres grip very well, wet or dry and they put a lot of rubber on the road. There's no doubt that the roadholding is above average and no one should get into too much trouble in an RS 2000 and even if he does, the car will respond to corrections favourably.

Ride is a bit too firm

While the handling is super, I cannot say the same thing of its ride. The standard Escorts ride remarkably well for a conventional design but in the interest of handling, Ford have sacrificed ride quality in the RS. At normal speeds, the car rides over normal bumps with jerks while smaller bumps are taken fairly well. At higher speeds, the ride improves and most irregularities are ironed out without too much drama, although I cannot say the ride quality at these speeds is as good as the Escort Ghia. At low speeds, the suspension does not seem to cope with even the

smallest imperfections and it becomes jiggly and can be rather uncomfortable.

Looking at it on the whole, it appears that while the ride is firm, it is also fairly well damped and except at low speeds, the ride works out to be not too uncomfortable. Strangely enough, when I found a stretch of bumpy, dusty little used road and drove it at speed, I actually found that the car could be fun to drive in such conditions. No wonder the Escort is so very successful in rallies!

The brakes are superb and so it should be in a performance orientated car. They are light and fade-free and with the sort of progressive action that is less likely to get anyone into trouble. In a panic stop, they stopped the car in a straight line and the wheels only locked after the car had been slowed down considerably. The handbrake is powerful and proved capable of holding the car on steep inclines.

Living with an RS 2000

Anyone who buys an RS 2000 is unlikely to be the sort of man who has no interest in cars whatsoever. After all, if it's family motoring he's after, the four door Escorts or the Cortina would do a better job. The RS is definitely for the sort of man who enjoys his motoring and is not too keen on compromis-

ing his love for fast driving and safety in a more ordinary saloon.

In Britain, you can buy similar performance for more money, but seldom for less. The home mechanic should find taking care of the car easy as the engine is well laid out with most of the everyday service items within easy reach. In addition, the car is surprisingly frugal on fuel and has enough power and handling to make it a very attractive proposition.

The boot is large and can take a fair amount of luggage so the car will serve as a good car for those week-end trips to the seaside.

It isn't perfect, of course. The ride is firm and rather uncompromising and interior accomodation isn't a strong point, but it has a lot more going for it than against it.

Will we see it here?

With the RS 2000, Ford have an excellent vehicle for the enthusiast as well as the sort of chap who might take the car to the track or rally circuit just to spectate. It is a delightful car to drive, has super instruments and contro)ls and more than enough power to satisfy the potential Harvey Yaps of the world.

I enjoyed driving Ford's top Escort and was quite reluctant to part with it after four days of rather enjoyable motoring. It's a good car and its success is well deserved.

If we are lucky, we may get our own version of the RS 2000 in Malaysia in the near future, but it will have to depend on whether there are enough 2 litre engines for Ford to supply to Ford Malaysia. If so, it shoud do Ford a tremendous amount of good as I'm sure almost everyone who now races or rallies in other machines may switch to the Ford product.

Test car : Courtesy of Ford of Europe Inc. Brentwood, Essex, England.

ROAD TEST

THE JUBILEE ESCORT

I'd be the last to admit that I've got a chauvinistic bone in my body (except where women are concerned), but there was something reassuring about getting behind the wheel of an European car after so many Japanese samples in succession. The immediate impression of tautness, good, firm ride and sombre interior trim was welcome enough to offset the obvious heaviness and not-so-exuberant engine. The reassurance was doubly telling because the test car was, on this occasion, no stranger. It's the evergreen Escort, and perhaps the most misunderstood variant at that, the 1.3GT.

But we've tried that before, haven't we? Sure we have, in October last year. However, there's a difference because this particular sample is one of 150 "limited edition" Anniversary Escort, complete with the embellishments and extras worthy of the name.

So let's call this article a re-test with some trimmings. Quite honestly, though, several aspects of the Escort came more clearly through on this the second time around, especially when no fewer than a dozen Japanese cars had given their share of impressions in the intervening period. If anything, they served to highlight the virtues of the test car more than ever.

This year is the 75th anniversary of Ford, but only the 14th of the Escort. For the former occasion the Ford people here have celebrated by doing up the ordinary 1.3GT and turning it into a collector's item. The Anniversary Escort comes with vinyl top, a nice black to offset the executive grey of the paintwork, and small but ostentatious items like front and rear spoiler, chrome wheel embellishers, fog lamps, series 70 radials, side strips and a quartz clock. And of course, things wouldn't be complete without the special Jubilee Badge stuck to the sides. All these extras would cost about a thousand-odd, and wonder of wonders! the car costs just about that much more (wholesale cost of items, that is) than the ordinary 1.3GT: $13,465 against $12,674 without insurance. We think the Ford people should have included an aircon unit, for what better

way for Malaysians to celebrate the marque's birthday than by keeping cool? Anyway, we think the price is more than fair because the obvious exclusiveness of owning a limited edition car is nil. Besides, Volvo resorted to a lucky draw where their Jubilee 264's were concerned, meaning that such exclusiveness could not be yours even if you had a million dollars. So we won't be at all surprised if by the time this article comes out, there won't be any Anniversary Escorts around to prove us wrong in our impressions.

Now that's all over with, let's get down to the car itself. For all intents and purposes the test car is an Escort 1.3GT. Now, that's nothing to sneeze at. This car forms one of the mainstays of Ford sales, an attractive package of the middle income family who prefers a European model over the myriads of Japanese lightweights. The usual price ($12,674.24) slots it right in the midst of cars in the 1200cc to 1600cc class. This is a safe place to be in because the 1300cc market is shared by only the BL models and that Hiroshima maverick, Mazda. The 323 has a following of its own at a slightly lower price than the 1.3GT, and so has the Allegro, at just $220 less. The other makes, Renault and Chrysler, are way behind to reckon strongly in the picture. It is against the 323 and the Allegro that the Escort 1.3GT has to show off its muscles.

And quite respectable muscles they are, too. The engine has given yeoman service (the marque's distractors would call it old), and in its present guise puts out a healthy 42kW, translating into a good power-to-weight ratio of 20kg/kW, better than either of its competitors. The 1298cc push-rod powerplant has a compression ratio of 9.2:1, necessitating the use of premium fuel, and is fed by a single Weber downdraught twin-barrel carburettor. Still, fuel consumption is on the frugal side, with in excess of 30mpg being realised under normal motoring conditions. Its on-the-road performance, too, does credit to the Ford name. Top speed is an impressive 150kph (we recorded 148kph indicated on this occasion) and acceleration an equally

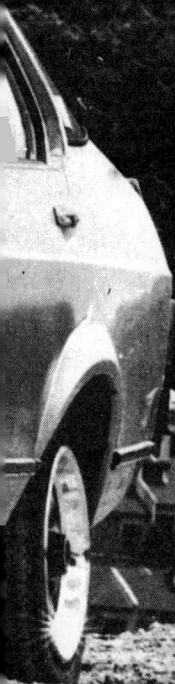

good 14.9 seconds average for the 0-100kph run (We obtained 15.2 seconds the last time, which may point to the positive contribution of the front spoiler).

So much for the mundane technicalities. On this the second time around, we were more struck by how strongly the Escort comes forth as a family car, and one with performance in reserve. Make no mistake, by "family car" we don't mean a sedate job. We mean a practical, respectably fast, four- to five-seater, a car that is easy on the pocket and offers few hassles in ownership. It is with this point of view that we began to appreciate how comfortable the Escort is, for both driver and passenger. My wife, who doubles up as the victim of our driving jaunts, commented favourably on the seats and the ride. Little wonder after all the jiggling (wash your mind out, reader!) she's had with the many Japanese cars.

Indeed, the virtues of the Escort 1.3GT must be, in strict order, its ride, handling, and performance. This is from the driver's point of view. To the passenger, the attractions must be the refinements of the interior and the comfort, which, compared to a Japanese car of the same class, are in the first instance rather restrained and in the second instance a revelation of how a car should feel. To the family man add the attractions of good fuel economy (nothing miraculous here unless you run on overinflated tyres at a steady 50kph) and easy serviceability.

We ran the 1.3GT like no normal family man would. Despite our sometimes deliberate hard use, the car acquitted itself well, thus making us wonder about how fickle car buyers can be, and how easily attracted they are to shiny knobs and pretty-pretty trim. This car, less nattily dressed in the Jubilee livery, is pretty remarkable in being able to hide its Hyde characteristics. This is a classic case of a full-blooded filly disguised as a family runabout, and if it serves well as the latter, it excells in the former role. Little wonder we see so many Escort in the rallies and GP's, for the handling is something else.

After so long with the Giesha group, I had almost forgotten how a car can be thrown about. The Escort is immensely chuckable, and the enthusiastic driver will soon catch on to the tactic of dashing into the apex of a corner, dabbing the brakes while simultaneously bringing the car round, and flicking back the steering while powering out. Guaranteed to make the tyres squeal and your passenger turn green. But it's all very safe, really, because the adhesion limits are high, especially with the low-profile radials. If you're lucky, you'll get an exciting three-point stance for a split second. We will be quite candid and say that this is something we'd not do with any rear-wheel drive Japanese car. The Escort inspires confidence not only in its actual ability to undertake this kind of manoeuvre, but also in its general stable "feel". This feeling of stability is miles (kilometres?) removed from the feeling of heaviness, though the two are easily mistaken. The Escort is endowed with a big-car feel, but moves with the agility of

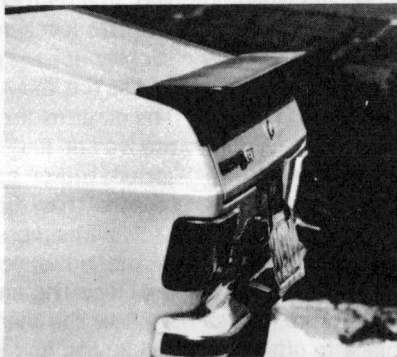

Right: For visual and maybe aerodynamic effects, you'll get the spoilers as standard. (Below): It could end up one day a collector's item, this anniversary tag.

a small Lancia, FWD notwithstanding.

The gearbox is easily the best around. Not surprising because Ford had many years to refine this contraption. The result is fast and precise gear changes, which, while without the oily smoothness of Japanese boxes, serve the driver well in his most delinquent driving. The slight aberration of the synchromesh which we found on the last occasion is now gone, pointing perhaps to slight variations in different samples. The throws are delightfully right, being neither too long or short, but we'd grumble about the shortness of the gearshift. An increase of about a half inch would make it perfect.

While the gearbox and gearshifts were found satisfactory, we were not too happy with the pedal layout. The distances between pedals are all right, but the accelerator pedal is way below the level of the other two. What is galling is the fact that this arrangement distracts from the otherwise beautiful relationship between car and driver. Heel-and-toeing is out, of course, and even the speedy

pivoting of the heel to tap the brakes. No doubt the enthusiast would remedy matters by putting an add-on Paddy Hopkirk pedal to the present pedal, so the shortcoming is not that serious.

Steering is by rack-and-pinion, an arrangement which explains the positive feel and the delightful precision of the action. Lock to lock turns amount to 3.5, just right for manoeuverability and lightness at parking speeds. This is one car that would not deter a woman driver, thanks to the high degree of driveability. The steering is light and precise enough, the gearshift smooth enough, and the pedal action almost effortless for the about-town puttering. Unisex, this car is, and not mundane in the bargain.

Visibility has always been a strong point of European cars, even at the expense of looking downright boxy. The Escort is, thankfully, possessed of presentable looks, and the Anniversary model carries a rakish, almost cocky, look, thanks to its spoilers and the vinyl trim. Still, front and rear visibility is good, the only blind spot being that sec-

tor occupied by the rear quarter pillars. There is, however, a serious omission to the package, and this is the absence of the wing mirror. Surely a ten-dollar wing mirror would not have amiss?

Another disappointing area to some might be the instrument panel. We would call it adequate for the car's pretentions, if plain. For practical purposes everything is there — tacho, speedo, fuel and temperature gauges, nicely clustered in a binnacle and fully visible from the driver's seat. Thus a complaint levelled against it would concern the aesthetic treatment. The all-black dash does very little to make the instrument panel appealing, and the retention of the chrome trim around the binnacle smacks of something from the 60's, not the 70's. A facelift, we feel, would be most welcome. Then there is the steering wheel itself — an odd affair with spokes central boss that bulge out, giving the driver that uneasy feeling that it'll be the first thing he'll encounter (a convex prod — nasty, neh?) in a collision. The controls and switches, however, are logically

Left: The chin spoiler gives the GT a bigger face and to decorate it further are a pair of Orimofoglamps. (Below): The shine in the wheel comes from the easily detachable embellisher.

THE JUBILEE ESCORT....

placed, with the indicator and horn controls incorporated in a stalk on the left of the steering column, and the light switches and wiper controls in two stalks on the right.

Ventilation is good without being superb like some of the Japanese. The blower works well enough, but could be better. Thankfully, among the anachronisms found in the Escort are the very welcome front quarter lights, and bringing in gobs of fresh K.L. smog into your face is no sweat (pardon!).

Night driving is served well by the headlights. The foglights are a welcome feature, good enough to teach hi-beam inconsiderates some manners. The headlights gave a good enough spread for steady 100kph driving, and the driver is moreover reassured by the better than average brakes.

Brake fade was evident after some hard stomps from quite high speeds, but this was not of a serious nature. The disc/drum system suits the car well, bearing in mind the medium weight (860kg) of the car and the speeds it is capable of achieving. We detected a certain sogginess in the pedal action, but will attribute this to the variations in this particular sample.

Back to the interior, and the extra, the Quartz clock. No doubt about it, it is super-accurate. But really, who'd care? The clock is located right where it matters least, right in the cavity ahead of the gear console, where your cigarette box and parking tickets should be. You'd need to bend down or at least peer through the steering wheel to tell the time. One consolation, however, you could always admire the legs of your companion on the pretext of looking at the clock!

The other additions found in the anniversary Escort may have doubtful value. We're willing to concede that the spoilers do in fact affect the performance of the car. After all, we did get better 0-100kph time than formerly. Besides, they do add to the aesthetic appeal of the body. The foglamps, well, they're welcome. As for the vinyl roof — to be quite honest, we're always been unenamoured of them because they make the interior pretty hot. Still, the snob value may make sweating it out quite worthwhile to some owners. An aircon would seem a

must, though certainly it will reduce the performance somewhat. The side strips are practical, while the wheel embellishers are merely decorative. The most valuable item to our minds — quite seriously! — is the Jubilee badge. Wait long enough and it'll be a collector's item. Watch it, though. Some kid can prise it off (there are two, actually, one on either side) without much trouble.

These mini-grouses and maxi-praises aside, the Escort, in Anniversary colours or everyday dress, remains an attracive package for the family man. The rear seats, for example, are almost sumptious, soft and supportive enough to rival the French saloon's. Legroom and headroom are generous. The boot, too, can easily take three super Samsonites, and is sensibly flat. There is logic, moreover, in the way the petrol tank is located, on the right. The weight of the petrol on the right is offset by the weight of the spare tyre, on the right. No digging out of luggage in the event of a puncture. For the owner who has frisky kids, the corded cloth-like material of the seats should be a welcome feature. It should last longer than the father's nerves.

We spent a happy three days with the Anniversary Escort, and were quite flattered by the curious looks of fellow drivers at traffic light. Of course the Escort couldn't match some of the nippier Nipponese jobs off the lights, but it didn't matter. We were more comfortably off, what with them jogging and jiggling all over the highway ripples. Besides, the spoilers gave us 'gaya'. During those three days we made a mental note to be kind to the next Nipponese job we test, because it's going to be tough taking its ride in our stride after the Escort. Maybe we should re-test an Opel Kadett first.

For the asking price, the Anniversary Escort is a whorthwhile buy. Sure, for the money one could get a go-faster jiggler, or a sedate fellow Britisher. But for a good all-rounder whose only tangible fault would be the rather restrained and old-fashioned interior, the Escort is hard to beat. Besides, for the money, where would you get a car whose front window winders worked the opposite way to the rear winders and to every other car we've tested? ●

JUBILEE ESCORT Specifications

Test Ratings

Key

★	poor
★ ★	below average
★ ★ ★	average
★ ★ ★ ★	good
★ ★ ★ ★ ★	excellent

Performance	★ ★ ★ ★
Economy	★ ★
Transmission	★ ★ ★ ★
Handling	★ ★ ★ ★ ★
Brakes	★ ★ ★ ★
Accommodation (luggage)	★ ★ ★ ★
Accommodation (passenger)	
Comfort	★ ★ ★ ★
Visibility	★ ★ ★ ★
Instruments	★ ★ ★ ★
Ventilation	★ ★ ★ ★
Noise	★ ★ ★ ★ ★
Finish	★ ★ ★ ★
Equipment	★ ★ ★ ★
Value for money	★ ★ ★ ★

MAKE	:	FORD
MODEL	:	Jubilee Escort 1.3 GT
COLOUR	:	Metallic Gray
PRICE	:	$13,465.61 (excluding insurance)
NO. OF FREE	:	1st, 500 miles
GUARANTEE	:	6 months or 6,000 miles whichever comes first.
ENGINE:		
Location	:	front
Cylinders	:	4, vertical, in-line
Valves	.	overhead, pushrods, and rockers
Carburettor	:	1 Weber downdraught twin barrel
Compression ratio	:	9.2:1

Bore & Stroke	:	81x63 mm
Cylinder head (material)	:	Cast iron
Engine block (material)	:	cast iron
Capacity	:	1298 cc
Maximum power	:	70 hp (DIN) at 5,500 rpm
Maximum torque	:	68 lb. ft. (DIN) at 4,000 rpm
Cooling system	:	water

TRANSMISSION:

Driving wheels	:	rear
Type	:	4-speed manual, fully synchronized
Clutch	:	s.d.p. (diaphragm)
Gear lever location	:	floor shift

RATIOS:

First	:	3.337
Second	:	1.995
Third	:	1.418
Fourth	:	1
Reverse	:	3.876
Final Drive	:	4.125

CHASSIS & RUNNING GEAR:

Construction	:	integral
Suspension, front	:	Independent, by McPherson, coil springs/ telescopic damper struts, anti-roll bar.
Suspension, rear	:	Rigid axle, semi-elliptic leafsprings, telescopic dampers, anti-roll-bar.
Steering type	:	rack and pinion
Lock to lock	:	3.5 turns
Steering wheel		
diameter	:	15''
Turning circle (between walls)	:	29.2 ft. (8.9 meters)
Brakes, type	:	disc/drum

DIMENSIONS (ins/mm):

Wheelbase	:	94.50/2400
Track, front	:	49.50/1257
Track, rear	:	50.60/1285
Overall length	:	156.80/3983
Overall width	:	61.80/1570
Overall height	:	54.50/1384
Ground clearance	:	4.92/125
Luggage capacity	:	15.4 cu ft.
Kerb weight	:	1896 lbs, 860 kg

REPLENISHMENT & LUBRICATION:

Engine sump capacity	:	5.8 pts
Engine oil, change interval	;	20w/50, every 6,000 miles
Gearbox capacity	:	1.6 pts
Final drive capacity	:	1.7 pts
Grease points	:	none

LIGHTING:

Headlamps	:	2
Battery	:	12V 38 Ah
Charging system	:	35A alternator

WHEELS & TYRES:

Wheels	:	5'', steel
Tyres	:	155 SR 13, Dunlop SP49.
Pressures—		
front	:	24 psi
rear	:	24 psi

FUEL CONSUMPTION:

As test	:	31 mpg
Touring	:	36 mpg
Fuel grade	:	Premium
Range	:	280 miles (approx)
Fuel tank capacity	:	9 gallons

PERFORMANCE (in kph)

Max. speed in gears

First	:	50
Second	:	85
Third	:	118
Fourth	:	150

ACCELERATION (in seconds)

Through the gears

0—40 kph	:	3.2
0—60 kph	:	6.2
0—80 kph	:	9.4
0—100 kph	:	14.9
0—120 kph	:	23.5

ACCELERATION:

In top gear

40—60 mph	:	6.0
60—80 mph	:	6.4
80—100 mph	:	7.5

STANDARD EQUIPMENT:

Adjustable steering
- Blower
Bootlight
- Carpets
Central locking
- Cigar lighter
- Clock
- Cloth trim (nylon type)
- Coat hooks
- Dipping mirror
Electric windows
Engine compartment light
- Fresh air vents
- Grab handles
- Hazard flashers
Head restraints
Heated rear window
Laminated windscreen
Map pocket
Glovebox
Outside mirror
- Parcel shelf front
Petrol filler lock
Front central armrest
Rear central armrest
- Rev counter
- Reversing lights
- Seat belts
- Seat recline
- Steering lock
- Sun visors
Tinted glass
- Tool kit
- Underseal
Vanity mirror
- Windscreen wash/wiper
- Wiper delay

WARNING LIGHTS

Handbrake
- Oil pressure
Door
- Charge
Low fuel level
Choke
Brake failure

Test car courtesy of: Universal Cars

FOR FORD, next weekend's Lombard RAC Rally will be a historic occasion. It may be the very last time in which a "works" Escort is ever entered in an international rally. The announcement has already been made that Ford are to withdraw (temporarily, it is said) from rallying, and it is thought that they will spend the next year or so developing a new rally car for the 1980s. Boreham, the base from which more than 100 rallying Escorts have emerged in the past 12 years, is dispersing its world-famous team of drivers. Next summer the Escort will be dropped from the production lines.

The RAC Rally, therefore, will signal the end of an era in rallying, which quite succinctly and correctly could be known as "the Escort years." No other rally car ever produced, in Britain or in any other country in the world,

The Escort years

For the past 12 years the Ford Escort has dominated the world of the international rally.
Now the end has come . . .

By Graham Robson

Top right: Ford had their first win with the Escort in April 1968, when Roger Clark and Jim Porter won the Circuit of Ireland

Right: In the 1973 Scottish, Escorts took the first 10 places. Chris Sclater, No. 4, finished sixth out of that bunch

has been used for so long, developed so continuously, or been even more competitive at the end of its life than at the beginning. For the irony of all this is that whereas the Escort Twin-Cam was off the pace by 1970, the Escort RS of 1979 is *the* car by which all performances are matched. The car which was no more than a strong "forest racer" in 1974, is now the most versatile rally car in the world. Its comfortable World Championship victory this year confirms it.

In 12 years, starting from the San Remo Rally of 1968, works-built Escorts have won almost every event open to them. They have been supreme

The Escort's first international appearance was in the San Remo Rally (March 1968), when Ove Andersson/John Davenport finished third overall.

on long-distance endurance rallies like the *Daily Mirror* World Cup marathon of 1970, fast enough and strong enough to win the Safari on two out of four factory-managed assaults, nimble enough and tractable enough to win on the snow and ice of Scandinavia, in the heat and dust of Greece, on the twisty tarmac of Portugal, in far-flung New Zealand, Canada, Hong Kong, or South Africa. Most of all, an Escort is still the supreme special-stage car.

And yet, for all that, the rallying Escort has made more comebacks than a temperamental opera singer. It was being beaten by rear-engined Alpine-Renaults and Porsches in 1970,

which led on the one hand to the birth of the RS1600, and on the other to the mid-engined GT70 project. Once again it was running out of pace in 1972 in 1.8-litre BDA-engined form, but was rescued by the alloy-block 2.0-litre version of that unit. It was looking old-fashioned in 1974, but rejuvenated by the new shape and new detail of the RS1800 in 1975. A return to World Championship rallies in 1976 showed its handling and technical deficiencies, but by 1978 the transformation and refinement was almost complete. Now, even in well-sponsored private hands, it may still be *the car* to beat.

The Escort Twin-Cam project was born at the beginning of 1967, when Ford's competition manager of the day, Henry Taylor, saw prototype Escorts running round the proving grounds at Boreham, sat down to conceive a "paper" car, and persuaded Walter Hayes (who controlled Ford's involvement in motor sport) to back him. On the basis that Boreham knew what they wanted, but that they would have to develop their own car, the idea was approved. The race to get a new car — a classic "homologation special" — into production was under way.

It would replace the Cortina-Lotus as Ford's fastest car, and to make the transition quick and easy it would also use most of that car's mechanical units — engine, gearbox, axle and front suspension. To get homologation as quickly as possible, it was decided to assemble the first 25 cars at Boreham, actually in the competition workshops, after

which day-to-day production would begin at Halewood, among the bread-and-butter Escort models.

The Escort range was announced in January 1968, and included a Twin-Cam prototype, but the first rallying appearance was in the Italian San Remo Rally, where Ove Andersson and John Davenport managed third

The Escort's first international win was in the Circuit of Ireland (April 1968), by Roger Clark/Jim Porter.

place overall, beaten by a Porsche 911 and a Lancia Fulvia HF. It proved two things — that the new car was fast enough, and that a lot of development was still needed. Bill Meade, who was Ford's rally engineer at the time, recalls that San Remo experience led to a crash programme to get a rear damper "turret" kit finalised, as the original (production-based) location was simply not up to the job.

There were no problems, at first, over performance. The well-proven Lotus-built engines,

slightly enlarged to 1,594 c.c. to bring them up to the limit of the 1.6-litre capacity class, could deliver a reliable 150 bhp in five-day rallying tune, the close-ratio Cortina-Lotus gearbox could withstand that sort of power, and the car was considerably lighter than its predecessor.

Roger Clark, too, found that it was not only lighter, but rather smaller, and considerably more precise in its road behaviour. He took an immediate liking to the Escort, and proved this by winning the first four events he ever did in the cars — the Circuit of Ireland, The Tulip, the Acropolis, and the Scottish rallies. *Autocar* borrowed the Scottish-winning car immediately after the event, found that it was geared right down with acceleration in mind, and that it could rush up to 100 mph from a standing start in no more than 21.2 seconds. It was dramatically better than anything the works-tuned Cortina-Lotus could achieve, and it was enough to make the Twin-Cam a potential winner in most events. It was in 1968, too, that two of the world's best drivers — Timo

The best of the 1.8-litre rallying BDAs pushed out about 205 bhp at 7,400 rpm, which was just enough to keep the Escort competitive, and it was certainly enough to propel Hannu Mikkola to a famous victory in the 1972 Safari. It was sweet revenge for Ford, who had sent a team of Twin-Cams to Kenya in 1971, only to be beaten for pace by the Datsun 240Zs which were their great rivals at the time.

It is now well-authenticated history that Brian Hart has designed a light-alloy version of the Ford cylinder block for racing-car use, that Ford's Peter Ashcroft literally tripped over the prototype block one day in Hart's workshops, and that it was speedily adopted for use in RS1600s. Hart has always intended that this new block should be capable of enlargement to the full two litres (he had formula 2 competition in mind), and this provided the final breakthrough which Ford needed. Roger Clark blooded the alloy block in the Jim Cark Rally of June 1972 (with a win on this national event), but really gave it a flying start in international competition by driving the 2.0-litre Escort (with Lucas fuel injection on this occasion) to its first RAC Rally victory in November, 1972.

Within months, every self-respecting Escort rally car, works or privately-sponsored, was using the alloy-block 2-litre, and it was at this time that the mechanical specification began to settle down. At the time it was company policy at the time that the Boreham team would concentrate on rough road events, rather than tarmac races, and it was almost inevitable that the specification of the cars began to get rather specialised. At the time, therefore, that Lancia were finalizing their mid-engined Stratos, and were about to dominate any tarmac event, especially where pre-event practice was allowed, Ford were turning more and more to building "forest racers", where strength and chuckability were more important than precise handling and outright speed.

The after effects of the Suez crisis of 1973-1974 didn't help.

Nor did the contraction of the team at the end of 1974, which resulted in Mikkola leaving, and in a more restricted programme taking place. The works cars, however, were still supreme in events like the British RAC, Scottish and Welsh rallies, in Scandinavia (whether on ice or snow, or in the dust of high

Makinen and Hannu Mikkola — first drove a works-prepared Escort, and both soon forged links with Boreham which were to last for a number of years.

But rallying, like motor racing, moves fast, and in 1969 there were three important events affecting Ford — Stuart Turner succeeded Henry Taylor as competition manager, the

Cosworth BDA engine was announced, and first details of the London-Mexico World Cup Rally were published. All in one way or another, were to be very important to the Escort's future.

Stuart Taylor, already famous for his work with BMC and the Mini-Coopers of 1962-1967, has a restless mind and is a great innovator, and arrived at Boreham just in time to prompt a great development programme for the Escort. The car's problem, already becoming clear in 1969, was that the Lotus engine

was near the end of its development, and that it was not going to be competitive for long.

This, and a testing programme connected with the 1970 World Cup Rally, resulted in strange Escorts with V6 engines and ZF gearboxes being used later in the year, and it also resulted in an important change in tyre contract — from Goodyear to Dunlop, who have served Ford well ever since then.

In a departate search for pace, Boreham produced a quartet of specialised ice-racers for the 1970 Monte Carlo Rally, complete with five-speed ZF gearboxes, Ford-Germany "Atlas" axles, four-wheel disc brakes and dry sump engines. Though this was, in effect, the birth of the modern Escort rally car, it was not enough for the cars to win this classic winter rally. Comprehensively beaten by three rear-engined Porsches and one rear-engined Alpine-Renault, Ford returned home with a determination to do even better. Stuart Turner and Roger Clark, it is said, conceived the bare bones of the mid-engined GT70 on the back of an enve-

lope, sketched out during the Nice-London flight after the Monte; Turner had already gained approval for the new Cosworth BDA engine to be fitted to the Escort, and for the new AVO factory to be set up at South Ockendon for the production cars to be assembled.

The GT70 project was abortive, not perhaps because of its design, but because of the 1971 Ford-strike which paralysed all production (and cash flow!) at the factory for so long in the spring of 1971, and because of

the difficulty of finding any place and any method in which to build replicas in small numbers. the RS1600 project, happily, was not.

Before this, however, Ford notched up their biggest victory, by completely dominating the *Daily Mirror* World Cup Rally with a massive team of 1.8-litre pushrod-engined cars. It was a complete vindication of the latest mechanical specification, and of Ford's meticulous practising and planning. The immediate result, too, was that the Escort Mexico was produced in considerable numbers, and was used by many private owners for their own motor sport.

Although the BDA engine, with its 16-valve cylinder head and Cosworth's accumulated experience built in to the breathing arrangements, was a lot more powerful than the Lotus unit it replaced, the cylinder block casting still restricted bore increases, and limited the engine capacity to 1.8-litres. It is worth noting, incidentally, that the RS1600's engine had been thoughtfully homologated at 1,601 c.c., which allowed such capacity increases.

summer), and in rough-road events all over the world.

At home, several frustrated performances in the RAC Rally (including the infamous year when all the cars suffered half-shaft breakages), had been succeeded by Roger Clark's 1972 win, and by the start of three successive victories for Timo Makinen. Ford's grip on the special problems of the RAC Rally were now such that they could build, service, and support a massive team for each event.

With the arrival of the new-shape RS1800, and an expansion of Boreham's activities for 1976, the work's team's programme began to include more far-flung events. In that year,

Waldegaard, won the Safari, the Acropolis and the Lombard-RAC rallies, while Kyosti Hamalainen won in Finland; in 1978 there were wins in Sweden (for Hannu

A works Escort won its first RAC Rally in 1972, driven by Roger Clark, and each RAC Rally since then has been won by an Escort.

Mikkola) and in the Lombard-RAC (for the seventh time — this time for Hannu Mikkola). Any number of seconds, thirds, and supporting performances by team drivers helped to confirm the fact that the Escort was back in world class in almost any conditions. Hannu Mikkola, you will

gaard's chances (by blocking the road on the penultimate special stage with rocks) he would probably have won the event. Now, the Monte is probably the only really important event which an Escort has never won.

In the past three seasons, and in spite of a tightening up of homologation regulations, the Escorts have become very sophisticated rally cars. At last there are tarmac, loose surface and "Safari tank" specifications, all of which show wide differences in body specification and particularly in detail suspension features. The BDA engine, now at the absolute peak of its development, can have carburettors or Kugelfischer fuel in-

Fettling, too, of a crippled car can be astonishing rapid. Complete gearbox swops take less than 15 minutes, struts take no more than five minutes a side, and even a cylinder head gasket (which is still something of a weak point on these 135 bhp/litre engines) can be swopped in 45 minutes. Given an hour in a service area, the space, light, and the spare parts with which to do the job, and Boreham's mechanics can produce the most astonishing improvements.

It is when one notices that differentials have been given oil coolers (mounted in the boot, and with electric fans used to draw air through them), and that the Panhard rod location has

---

Below: Bjorn Waldegaard and Hans Thorszelius head for deep water during the first leg of the 1977 Silver Jubilee Safari, which they won convincingly

1968 "works" car: Escort Twin Cam

Engine: 4-cyl, in line, in five-bearing cast-iron cylinder block; modified by Lotus from Ford Cortina 1500 base. Bore, stroke and capacity 83.5x72.8 mm, 1,594 c.c. Light alloy cylinder head, with two valves per cylinder, and twin overhead camshafts driven by roller chain. Twin side-draught dual-choke Weber carburettors. Approximately 150 bhp at 7,200 rpm.

Transmission: Single dry plate clutch and Ford four-speed all-synchromesh gearbox. Remote control gearchange. Hypoid bevel live rear axle with limited-slip differential.

Chassis: Unit-construction pressed-steel two-door saloon body/chassis unit. Independent front suspension by MacPherson struts, coil springs and anti-roll bar. Rack-and-pinion steering. Rear suspension by half-elliptic leaf springs and twin radius arms. Front wheel disc brakes, drum rears. Cable operated handbrake. 13in. cast-alloy Minilite wheels, with choice of Goodyear tyres.

1979 "works" car: Escort RS

Engine: 4-cyl, in line, in five-bearing cast light alloy cylinder block, derived from Ford Cortina 1600 base by Brian Hart Ltd. Bore, stroke and capacity 90.5x77.62 mm, 1,998 c.c. Light alloy cylinder head, with four valves per cylinder, and twin overhead camshafts driven by cogged belt. Twin side-draught dual-choke Weber carburettors, Lucas or Kugelfischer fuel injection, depending on application. Between 250 bhp and 270 bhp at 8,500 rpm. Safe limit, 9,500 rpm.

Transmission: Triple or single dry plate clutch and ZF five-speed all-synchromesh gearbox. Remote control gear change, choice of extension lengths. Hypoid bevel line rear axle with ZF limited slip differential.

Chassis: Unit-construction pressed-steel two-door saloon body/chassis unit. Independent front suspension by MacPherson or modified-type struts, coil springs and anti-roll bar; tension struts in certain specifications. Rack and pinion steering. Rear suspension by half-elliptic leaf springs (helped by concentric coil spring/damper units in some specifications), four radius arms, and a Panhard rod. Four wheel disc brakes. Cable and hydraulically operated handbrake. 13in. or 15in. cast-alloy Minilite wheels, with wide choice of Dunlop tyres.

Finn Kyosti Hamalainen seems to be equally at home in his native land during the 1977 1000 Lakes, centre, or in the Scottish Highlands on the 1978 Scottish, far right

note, had re-joined the team at the start of 1978.

In Britain there was little doubt that the Escorts were still the standard by which other cars were to be judged. Vauxhall produced their Chevettes, which were remarkably similar in mechanical layout, while Chrysler (Talbot, now) had one false start with the Avenger-BRM but a much more convincing try with the Sunbeam Lotus of 1979. Roger Clark's dominance of the scene was challenged by Russell Brookes (who won the RAC Rally Championship in 1977), and overshadowed by a series of virtuoso Mikkola performances in 1978.

Development at Boreham, guided by Alan Wilkinson, but directed by Peter Ashcroft, had refined the Escort to a remarkable degree by the beginning of 1979, and the cars entered for the Monte Carlo rally were as specialized as any Stratos or Fiat could possibly be. The record shows that the attempt was a failure, that a so-called "private" Stratos beat the works Escorts; the fact is that without the apparent sabotage of Walde-

jection, depending on the application, and the Monte cars were reputed to have nearly 270 bhp in ultimate "tarmac race" tune. There are so many different gearbox, axle, wheel and tyre combinations that even the staff at Boreham have to refer to their charts.

And yet the BDA engines we see in use on the British-specification "forest" cars idle smoothly, pull gently in traffic, and have an astonishingly progressive torque curve. Peaking at 8,500 rpm, they can nevertheless be revved to 9,500 rpm if the need arises (and, in the heat of the moment, it usually does!).

been reversed to allow more space for the exhaust system run, and that roll cage layouts were revised even during 1979, that one realises how much work was still being put into the Escort as a competition car.

Now, at the end of 1979, after a phenomenally successful 12-year career, the last "works" Escorts have been built, and for the next couple of years at least, the private Escorts will once again come into their own.

Will Boreham's successor, in whatever form it may take, be anything like as competitive? Will it win Safaris and tarmac events? In winter or summer? Ford hope so.

---

however, Timo Makinen's win in the South African Total, and his second place in the Spanish Firestone event, have to be judged against team failures in Morocco and in Australia.

By the end of 1976, however, Bjorn Waldegaard had joined the Ford team after being inexplicably sacked by Lancia, and a serious and well-planned development programme was being mapped out for the cars. Neither in 1977, nor in 1978, did Ford make a whole-hearted attempt on the rallying World Championship, but this did not stop them indulging in face-to-face battles with the Fiat Abarth 131s. In 1977 the new recruit,

Making merry

Four fun cars that will bring a smile to even the most jaded drivers

FORD has got a great deal to answer for! For the Dagenham concern has done more for the enthusiast driver — the so-called 'Boy Racer' — than any other firm. Since the demise of the British Racing Green sports car with its soft top but rock hard ride, Ford has been responsible for providing saloon cars that will out-corner and out-perform what few two seaters there are left, as well as provide a reasonable amount of space, and at a price within the grasp of many.

Once, the go-faster element drove Mini-Coopers while a few struggled with the temperamental Lotus Cortinas, but since the advent of the Escort the tide has turned Ford's way. The enthusiasts have been queuing up to drive Escort Mexicos, RS1600s, RS1800s and latterly RS2000s. And it wasn't too long ago that Ford had this corner of the market to themselves — there was the up-market Triumph Dolomite Sprint and the special order left hand drive Opel Kadett GT/E, but that was about it.

And then, earlier this year, came a horde of go-faster specials veritably tripping over each other to attract would-be buyers . . . fuel crisis or no. Some of the manufacturers who set about turning their bread and butter cars into hot shoe shufflers had tried — half heartedly — before. Vauxhall once gave us the Firenza Droopsnoot, now they hide the

Engine

Ford

For
Lively unit; excellent gearbox; easy to maintain

Against
Noisy and rough; hard driving will mean poor economy

Talbot

For
Powerful; proven reliability; electronic ignition

Against
Lumpy, rough and temperamental in town; thirsty; unrefined

Vauxhall

For
Technically interesting; powerful and smooth

Against
Fouls plugs in town; thirsty; expensive

Volkswagen

For
Smooth; powerful; economical; good gearbox

Against
Buzzy at high speeds

Chevette 2300HS under a bushel; Talbot, née Chrysler, née Hillman developed, if that's the right word, the Avenger Tiger before using the same formula for the Talbot Sunbeam 'Ti; Renault hotted up their baby 5 calling it the Gordini in this country; and Volkswagen fuel injected the Golf dubbing it the GTi. The final manufacturer to enter the fray was Fiat who did a double, by squeezing 70 bhp out of their 1050 cc 127, as well as dropping a two-litre engine into a 131 shell and calling it the Mirafiori Sport.

For this test we have chosen three of the newest examples and compared them with the Daddy of them all, the RS Escort. As it turns out engine sizes and prices of the quartet vary quite considerably, but all appeal to the same sort of buyer and all provide the same sort of fun. And — as we shall see — all offer much the same performance despite it all.

The cars

Leading the group is the elder statesman, the RS2000 Custom at £5181. We did in fact test one of these cars earlier in the year (**What Car?** March '79) but as outlined in that test we were not happy with the performance figures we achieved at a streaming wet and slippery test track. Neither were Ford. This was a good excuse to put the car through the test routine once more.

The rivals chosen, all hatchbacks, are the Talbot Sunbeam Ti, the cheapest at £4528; the Vauxhall Chevette 2300HS, at £5939 the most expensive of the group; while in the middle comes the newest, the £5010 Volkswagen Golf GTi. We had hoped to also put a Renault 5 Gordini through the test routine but Renault's test fleet examples are in constant demand and so one was unavailable at the time of the test . . . another time, perhaps.

Performance

Although all four cars have four cylinder engines, the sizes of those engines vary considerably. There are a pair of 1600cc cars — the Talbot and Volkswagen — the Ford 2-litre, and topping the pile, the 2.3 litres of the Vauxhall. And yet quoted engine powers are close. Smallest is the Sunbeam with a maximum 100 bhp, the

Ford and Volkswagen both develop 110 bhp, while the Vauxhall again comes top with 135 bhp.

It's the same story on the track where performance times against the clock are uncannily similar. It's little use having masses of power if the car has to waste part of that power dragging a heavy body along, thus it should come as no surprise that the lightest car by some 2 cwt — the Golf GTi — is the quickest accelerating. From a standing start to 60 mph the German car takes just 9.2 seconds, which is no bad time for a 1600 cc car. The most powerful, and heaviest the Chevette is next quickest taking 9.7 seconds, while the Ford and Talbot both take 9.9 seconds. As we said the cars all perform in much the same way as far as the stopwatches are concerned, up until mid speeds at least. The power advantage of the Chevette becomes apparent at higher speeds, though, being the quickest above 70 mph by some margin.

Over the standing 400 metres it is the Vauxhall and Volkswagen that show the others the way home, both taking 17.3 seconds, though the Vauxhall has a far higher terminal speed indicating that while the Golf has begun to run out of steam, the Chevette has just got the bit between its teeth. It should be said at this point that, once again, we were not happy with the performance of the Escort RS2000. As fate would have it, the car tested in this issue is exactly the same car as we tested in the March issue, and without the hindrance of the poor weather we improved our 0-60 mph performance times from 10.4 seconds to 9.9 seconds; Ford quote a time of 8.5 seconds. Short of blowing up an engine, we decided our times were as good as we were going to record and called it a day, deciding either Ford's claim or BPU 917T was at fault. In any case 9.9 seconds to reach 60 mph from rest is not a time to be scorned.

While the typical boy-racer might consider standing start acceleration times the most important aspect of a car's performance, we believe that more significant are top gear overtaking times. And here the Escort comes out on top, accelerating from 50-

TALBOT SUNBEAM Ti, VAUXHALL CHEVETTE 2300 HS, VOLKSWAGEN GOLF GTi

70 mph in top in 8.9 seconds, compared with the Talbot's 9.3 seconds, the Chevette's 9.9 and the Golf's 10.4 seconds.

The Vauxhall is the only one of the group with a five speed gearbox, though in this instance direct comparisons are permissable because the Chevette's is a close ratio five speed 'box rather than the more normal four speed plus overdrive ratio five speed gearbox.

The most important aspect of performance is how the car behaves on the road, and in this case the story takes a distinct twist for two of the foursome are built primarily as club racing specials — or that is how it seems when trying to thread the Vauxhall or Talbot through busy city traffic. Both the Chevette and Sunbeam with their twin carburettors, seem to spend the bulk of their time fouling their plugs and popping and banging their way through anything that isn't a flat out blind.

To enter cars in various forms of racing or rallying, it is necessary to prove to the authorities the car is a standard road car available to all and sundry, thus some manufacturers make the barest minimum needed. We feel the 16-valve twin overhead camshaft engine of the Chevette and the twin Weber version of the Sunbeam 1.6 litre engine make both these car homologation specials — ideal for special stages and circuit racing, but out of their depth in normal traffic.

Ford are past masters of the homologation special, too. But with the RS2000 they have played safe, for the car is basically an Escort with a 2-litre Cortina engine under the bonnet and behaves in traffic in much the same way a Cortina does. Volkswagen's answer is by far the most successful. Rather than inserting larger engines or playing with temperamental twin carburettors that need expert tuning, the Golf engine has been fuel injected allowing a power boost with none of the problems faced by the rivals. Of course fuel injection is not cheap, nor did it have a good realiability record, but the Germans have been injecting engines for years and have by all accounts got it right now.

On the road it is the Golf that feels the smoothest to drive, with its delightfully free-revving attitude, the 1588 cc developing peak power at a high 6100 rpm. The Escort and Chevette both feel powerful but lack the ability to rev freely like the Golf while the pushrod Sunbeam unit feels simply archaic. For once gearchanges on all the cars can receive praise rather than rebukes, though as usual that in the Escort is the best. The small gate and an ideally placed gearlever which snicks in and out of gear means the driver is forever changing gear just for the fun of it. Changes in the other three are also good, though the Chevette, which has an unusual gear change with the top four ratios in the usual H pattern and first on its own to the left, suffers a heavy clutch.

It is usual to find speedometers a touch on the optimistic side — it probably saves quite a few licences, but the speedometer on the Escort was hopelessly optimistic while that on the Chevette under-read marginally.

Handling

With four such cars you would be forgiven for thinking that handling and roadholding capabilities should be of the highest order. You would not be disappointed, though there is one distinct leader and one distinct loser in this group. Out at the front must be the only front wheel drive car of the foursome. We have, in the past, heaped praise on the handling of all Golfs, whether 1100 cc, 1500 cc or diesel variants and the same must go for the GTi. It shares the same basic suspension set-up as other Golfs, but set slightly lower and slightly wider thanks partly to the use of fatter tyres, and the result is an agile all-rounder that corners with mild understeer but it is more than willing to help a driver hurry by lifting an inside rear wheel when asked.

Traditionally, rear wheel drive has been demanded by the enthusiast simply because it is easy to slide the rear round a corner by some unsubtle work with the right throttle foot. The result is a car that goes around corners quickly, but untidily, and a car that can catch out the unwary.

Opposite locking one's way along is fine on a test track, but it is not at home on public roads. For that reason we like the Golf so

Interior

Ford
For
Excellent Recaro seats; superb driving position
Against
Rear access poor; limited rear kneeroom

Talbot
For
Well shaped seats; hatchback
Against
Poor rearward movement of front seats; limited space in rear; sombre

Vauxhall
For
Good driving position; cheerful interior; well trimmed
Against
Poor rear space; access to rear difficult

Volkswagen
For
Firm, but comfortable seats; bright interior; good driving position
Against
Cheap finish; uninspired dash; engine noise intrudes

Instruments

Ford Escort RS 2000

1: Clock
2: Two-speed fan
3: Heater controls
4: Rev counter
5: Fuel gauge
6: Temperature
7: Oil pressure
8: Speedometer
9: Air vent
10: Lights master switch
11: Two-speed wipers
12: Indicators/dip/flash/horn
13: Rear fog lamp
14: Heated rear window
15: Cigarette lighter
16: Hazard warning lights

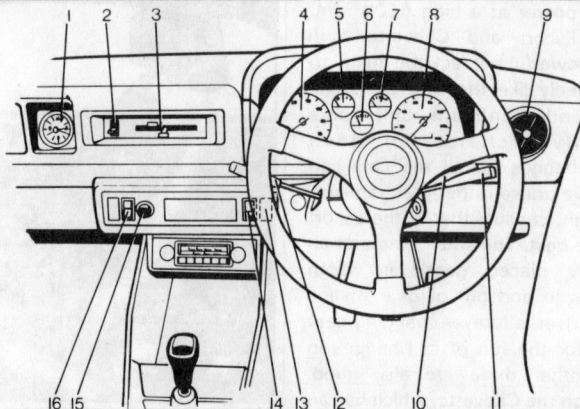

Talbot Sunbeam Ti

1: Front fog lights
2: Heater controls
3: Two-speed fan
4: Spare
5: Indicators
6: Lights/dip/flash/horn
7: Temperature
8: Oil pressure
9: Rev counter
10: Speedometer
11: Fuel gauge
12: Battery condition
13: Two-speed wipers/washers
14: Heated rear window
15: Choke
16: Rear fog lights
17: Brake test
18: Hazard warning lights
19: Rear wash wipe

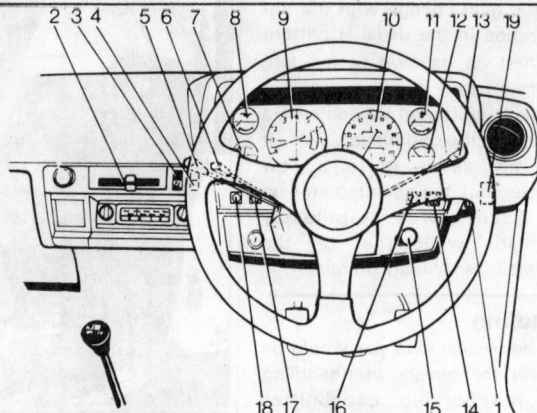

Vauxhall Chevette 2300HS

1: Two-speed wipers/washers
2: Speedometer
3: Clock
4: Rev counter
5: Lights master switch
6: Indicators/dip/flash/horn
7: Hazard flashers
8: Choke
9: Interior light
10: Heated rear window
11: Cigarette lighter
12: Voltmeter
13: Temperature gauge
14: Oil pressure gauge
15: Fuel gauge
16: Two-speed fan
17: Heater controls

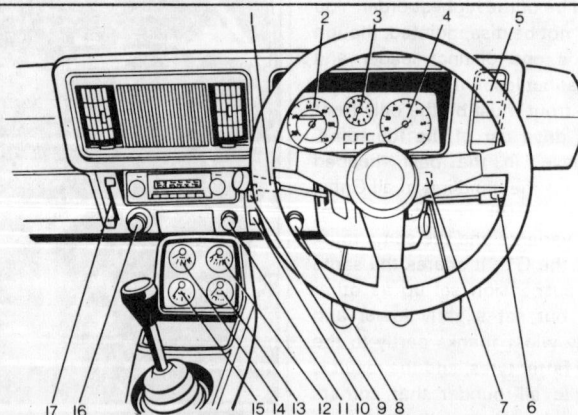

Volkswagen Golf GTi

1: Cigarette lighter
2: Three-speed heater fan
3: Hazard warning lights
4: Heated rear window
5: Temperature gauge
6: Rev counter
7: Fuel gauge
8: Speedometer
9: Panel light rheostat
10: Lights master switch
11: Two-speed wipers/intermittent rear wash-wipe
12: Horn
13: Indicators/dip/flash
14: Heater controls
15: Clock
16: Oil temperature

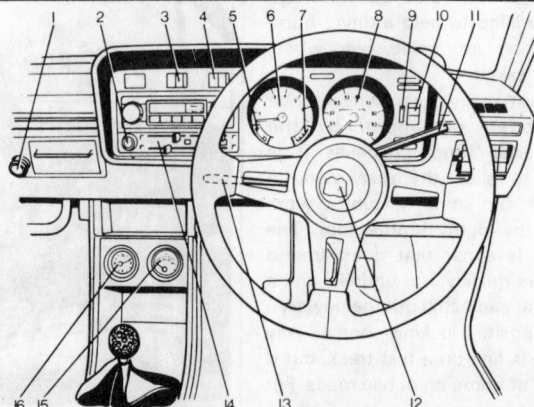

much, but that is not to say the Escort and Chevette do not handle well, for they do. The Escort with its rather crude rear semi-elliptic suspension nevertheless has such delightfully light and precise steering that the whole car feels well balanced, while the live-rear axled Chevette with its wide, fat tyres is difficult to unstick in the dry at least. The Vauxhall's steering — all have rack and pinion steering — is not as light as the Escort's but precise and quick, despite it having four turns lock-to-lock.

At the back comes the Sunbeam. We consider the ordinary shopping Sunbeams to be safe if a trifle dull handlers, and were looking forward to the Ti in the hope that Talbot have given th car a little more agility. They haven't. It still feels stodgy and unexciting.

As all four cars are aimed at the enthusiast, ride comfort has been sacrificed to some degree on the altar of handling. But once more it is the little Golf that comes out on top with a ride that is firm, in a typical Germanic way, but in no way uncomfortable. The car feels stiff and taut and yet is ideal transport for long journeys, while the Escort driver would soon get pretty fed up with the continual crashing and bumping as the tight suspension seemed to find all the pot-holes available. The Chevette and Sunbeam both come over as rally specials with harsh rides. Fortunately all four have excellent front seats which hold the driver and passenger in place during even the most spirited driving, especially so the Recaro seats found in the Escort.

The foursome also share the same braking arrangement with servo assisted discs at the front and drums at the rear and — thankfully — all complete their task with contemptuous ease, though the Golf has the usual VW sponginess.

Accommodation

It can be argued that these modern sports cars do not have to be particularly spacious, but even boy-racers have to do the shopping some times!

Obviously with cars that are variations on bread and butter saloons and hatchbacks, it stands to reason that they will perform mundane tasks as well — or as

WHAT CAR? CQMPARES RS2000, SUNBEAM Ti, CHEVETTE HS, GOLF GTi

Ford Escort RS2000 Custom

Familiar lines of the Escort with added RS spoiler. Boot (below) is small and crowded

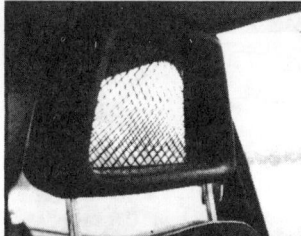

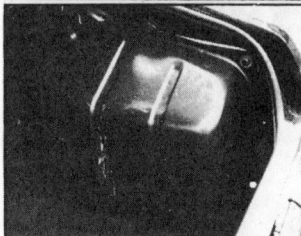

Top: controversial tennis racket head rests
Centre: rubber droop snoot disguises Ford
Above: the boot is full of petrol tank

Talbot Sunbeam Ti

Stick-on stripes and spoilers mark the Ti. Hatchback (below) is on small side

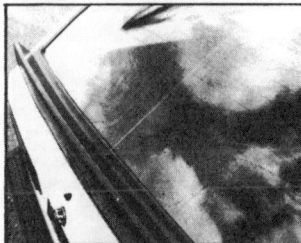

Top: washer is directed through spoiler
Centre: twin Webers provide extra power
Above: stripes come unstuck with petrol

badly — as their cheaper relatives. All four have but two side doors, which means that getting to and from the rear of the cars can be a problem. And once there, room is limited anyway — especially in the three home grown cars. The Golf at least has a reasonable amount of rear leg and head room. In the front the story is not so bad, though taller drivers will find themselves at a disadvantage in the Chevette and Sunbeam as in order to allow those in the rear some room, the front seats do not adjust back far enough — and those special seats take up even more of the available space than the usual seats.

Stowage space in the rear of the cars is reasonable if not outstanding. The smallish boot of the Escort loses space to the spare wheel and petrol tank, while the boot floors of the Sunbeam and Chevette are none too deep thanks to the stowing of spare wheels underneath and the restrictions of rear wheel drive. Back seats in the three hatchbacks all fold down, however, increasing space further. The boot area of the Golf is rather narrow but much deeper than its rivals. Unlike the top of the range pedestrian Sunbeams, the Ti does not have a split rear seat arrangement which would make the car more versatile.

Interior stowage space is nothing special in any of the cars — three have small glove boxes, while the Sunbeam has none. Vauxhall and Volkswagen have given their cars a centre console with small shelves for odds and ends, while Talbot have taken a centre console *out* of the Sunbeam . . . the GLS has one, the more expensive Ti does not.

Living with the cars

The major bugbear of any tuned car is noise, and it is a problem faced by three of these four, the exception being — yet again — the civilised Golf. In contrast to the others, the Golf feels like a limousine with little wind or tyre noise, though there is a reminder that this is no ordinary Golf from the excited note of the exhaust and engine. The Ford suffers from engine noise while the most predominant noise in the Talbot is the typical asthmatic breathing of twin Webers. The Vauxhall is just plain noisy, the exhaust note is loud, the

Wide wheels and spoilers give Chevette a mean look. Hatchback (below) is useful

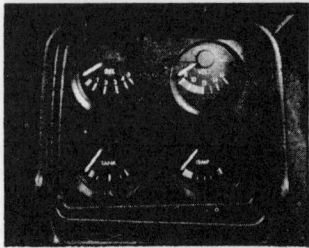

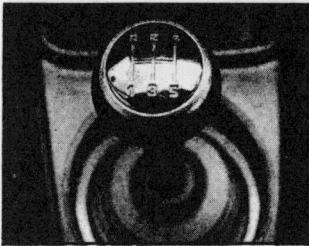

Top: brutal nose transforms car's looks
Centre: gauges in console are poorly placed
Above: unusual five-speed gearbox pattern

Golf GTi is understated — just one stripe. Boot (below) spoilt by intrusions

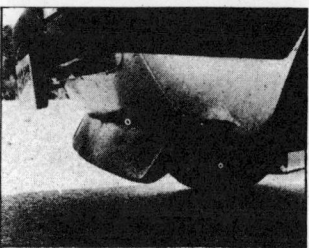

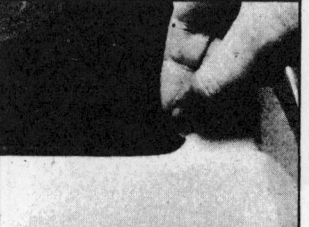

Top: nose spoiler helps stability at speed
Centre: A visual pun — a golf ball gear knob
Above: matt black 'paint' is really tape

wide tyres rumble and the engine makes sure everyone inside knows just how hard it is working.

Driving positions are generally excellent — especially so in the Escort, but none is perfect. The Vauxhall has a cluster of four dials sitting in a floor mounted console and are virtually unreadable; the Sunbeam needs more rearward adjustment to the seat, and has three spindly stalks to operate the major controls, which is one too many. The latter problem also affects the RS2000, while the Golf's floor mounted pair of dials are difficult to read at a glance — these are minor niggles however.

All four have spares stowed in the boot, upright in the case of the Ford, but beneath all the luggage in the remaining trio. All four have high loading sills, too.

Ignoring the problems of town driving as outlined in the Performance section, there could be further problems living with the Vauxhall and Talbot, and to a lesser extent with the Ford, for they all advertise the fact that they are boy-racers and attract attention to themselves. The Chevette has a neatly moulded air dam and masses of go faster stripes proclaiming its potential, while the Talbot uses black stick on stripes as if they were going out of style. The Talbot also has a air dam which in this case houses a useful pair of fog lamps. Ford designers have managed to turn a rather ordinary looking saloon into something special without recourse to stickers and stripes, but the rubber boot spoiler and "droopsnoot" nose of the RS2000 are rather obvious.

Volkswagen, on the other hand, have left the Golf too standard for some tastes. Externally the major differences are a red rim to the grille, and a matt black surround to the rear window and that's about it. Volkswagen themselves say it's case of "understatement instead of warpaint" and it's an image **What Car?** believes is probably the right one.

Heating and ventilation has come a long way in recent years since the advent of through flow ventilation and only the Sunbeam disappoints in this respect — there just doesn't seem to be enough power through the face level vents to do the job properly. Only the Vauxhall has opening rear

WHAT CAR? COMPARES/RS2000, SUNBEAM TI, CHEVETTE 2300 HS, GOLF GTI

Car	Ford Escort RS2000	Talbot Sunbeam Ti	Vauxhall Chevette	Volkswagen Golf GTi
Price	£5181	£4433	£5939	£5135
Performance				
Max Speed (mph)	107	111	115	108
Max in 4th (mph)	—	—	93	—
Max in 3rd (mph)	90	76	72	90
Max in 2nd (mph)	63	50	51	61
Max in 1st (mph)	34	30	32	34
0-30 (sec)	3.7	3.1	3.4	3.5
0-40 (sec)	5.6	5.2	5.2	4.8
0-50 (sec)	7.4	7.0	7.1	6.9
0-60 (sec)	9.9	9.9	9.7	9.2
0-70 (sec)	13.5	13.2	12.3	12.7
0-80 (sec)	17.6	16.9	15.8	16.6
0-90 (sec)	24.1	23.3	20.1	21.6
0-400 metres (sec)	17.8	18.0	17.3	17.3
Terminal speed (mph)	81	82	87	82
30-50 in 3rd/4th/5th (sec)	8.7	8.9	7.7/12.2	9.7
40-60 in 3rd/4th/5th (sec)	8.5	8.4	7.3/9.5	9.6
50-70 in 3rd/4th/5th (sec)	8.9	9.3	6.7/9.9	10.4
Speedo error at 60 mph	12.5% fast	6% fast	1% slow	3% fast
Specifications				
Cylinders/cap (cc)	4/1993	4/1598	4/2279	4/1588
Bore x stroke (mm)	90.8 x 76.9	87.3 x 66.7	97.8 x 76.2	79.5 x 80
Valve gear	ohc	ohv	dohc	ohc
Main bearings	5	5	5	5
Power/rpm (bhp)	110/5500	100/6000	135/5500	110/6100
Torque/rpm (lbs/ft)	117.6/4000	96/4600	134/4500	96.5/5000
Steering	rack/pin	rack/pin	rack/pin	rack/pin
Turns lock to lock	3.5	3.7	4.0	3.3
Turning circle (ft)	32.8	31.5	30.2	32.8
Brakes	S/Di/Dr	S/Di/Dr	S/Di/Dr	S/Di/Dr
Suspension front	I/McP/A	I/McP/A	I/Wi/C/A	I/McP
rear	½E/Ra	4L/C	C/A	I/C/TCA
Costs				
Test mpg	19.3-23.0	18.0-19.7	15.3-18.1	23.0-30.4
Govt mpg City/56/75	25.7/37.1/27.6	21.9/34.3/00.0	17.5/34.4/28.0	23.0/41.5/32.5
Tank galls (grade)	9(4)	9.0(4)	8.4(2)	9.9(4)
Major service miles (hours)	12,000 (4.45)	10,000 (2.17)	6000 (3.6)	10,000 (2.4)
Parts costs (fitting hours)				
Front wing	£28.66 (—)	£20.00 (—)	£27.85 (7.9)	£37.41 (1.6)
Front bumper	£17.00 (—)	£23.60 (0.48)	N/A	£38.77 (0.4)
Headlamp unit	£24.47 (0.5)	£16.97 (0.5)	£23.00 (0.3)	£15.25 (0.5)
Rear light lens	£3.96 (—)	£10.10 (0.17)	£4.94 (0.2)	£11.78 (0.4)
Front brake pads	£12.88 (0.8)	£14.60 (0.53)	£14.40 (0.7)	£9.38 (0.7)
Shock absorber	£11.11 (0.8)	£14.61 (0.8)	£23.90 (0.6)	£34.07 (0.8)
Windscreen	£30.00 (1.5)	£45.00 (1.02)	£48.85 (1.9)	£33.95 (0.7)
Exhaust system	£83.90 (0.8)	£69.93 (0.58)	£74.67 (3.1)	£79.63 (1.3)
Clutch unit	£38.03 (2.4)	£34.39 (1.9)	£29.72 (8.0)	£36.76 (0.7)
Alternator	£88.11 (0.5)	£37.70 (0.35)	£35.65 (0.5)	£41.75 (0.9)
Insurance group	7/on app	4/£80-101	6/£118-133	6/£118-133
Warranty/quote	12/UL	12/UL	12/UL	12/UL + 6yr anti-corrosion
Equipment				
Alloy wheels	Yes	Yes	Yes	Yes*
Five-speed gearbox	No	No	Yes	Yes*
Electronic ignition	No	Yes	No	No
Clock	Yes	No	Yes	Yes
Front fog lights	No	Yes	No	£95.14
Rear fog lights	No	£20.22	No	One
Laminated screen	Yes	£46.13	Yes	Yes
Tinted glass	Yes	No	Yes	£109.34
Ammeter	Yes	Yes	Yes	No
Oil temperature gauge	No	No	No	Yes
Petrol cap lock	No	No	No	Yes
Radio	No	No	Yes	No
Intermittent wipe	Yes	Yes	No	Yes
Rear wash-wipe	N/A	Yes	No	Yes
*1980 models				
Dimensions				
Front headrooms (ins)	38	35	36	36.5
Front legroom (ins)	36-41	33-39.5	31-37.5	31.5-39
Steering-wheel-seat (ins)	13-19.5	11-17.5	12-18	12-19.5
Rear headroom (ins)	32	31.5	32	34
Rear kneeroom (ins)	23-29	22-29	23-29	25.5-32
Length (ins)	161.8	150.7	157.2	150.2
Wheelbase (ins)	94	95	94.2	94.5
Height (ins)	55	54.9	53.6	54.7
Boot load height (ins)	30	34	32.5	30
Boot depth (ins)	39	25-46	31-51	30-43
Boot height (ins)	21	10-28	6-24	16-39
Boot width (ins)	40-46	40-51	31-49	35-51
Overall width (ins)	61.6	63.1	62.2	64.0
Track F/R (ins)	50.8/51.8	51.8/51.3	53.4/53.2	55.3/54.0
Int width F/R (ins)	50/50	53.5/52	51/50	52/51
Weight (cwt)	18.2	17.9	18.9	15.9
Towing weight (cwt)	19.7	15.7	N/A	23.6
Payloads (lbs)	1000	970	783	926
Boot capacity (cu ft)	10.3	14.7/42.7	9/35	13/38.9

KEY. Valve gear: ohc, overhead camshaft; ohv, overhead valve; dohc, double overhead camshaft; cih, cam-in-head. **Steering:** rack/pin, rack and pinion; rec ball, recirculating ball; worm/nut; worm/roll, worm and roller; PA, power assistance. **Brakes:** Di, discs; Dr, Drums; S, servo assistance; P, power assistance. **Suspension:** I, independent; C, coil springs; Tor, torsion bar springs; ½E, semi-elliptic springs; Hg, Hydragas; Hp, Hydropneumatic; Wi, wishbones; McP, MacPherson struts; deD, de Dion beam; Ta, trailing arm location; STA, semi-trailing arm location; 4L, four link location; 3L, three link location; TL, transverse link; DA, dead beam axle; WL, Watts linkage; PR, Panhard rod; TCA, torsion crank axle; RA, radius arms.

windows to aid air extraction, a feature that is sorely missed in the others.

Equipment

Talbot seem determined the Ti should be regarded as a club special by basing the model on the cheapest model in the range, presumably thinking the rally driver is not bothered by the lack of a cigar lighter, panel rheostat, luggage tonneau cover and split rear seat. But as a regular road car these omissions add up and give the drive the impression the car has been put together on the cheap. Vauxhall on the other hand have made sure the HS is as comfortable as possible. It has a laminated screen, tinted glass and a radio as standard, as well as alloy wheels and a five speed gearbox. The Talbot does have alloy wheels — well four of them, as the spare is a normal steel wheel.

RS2000s are available in two guises — this is the more expensive Custom version which features alloy wheels, those excellent Recaro seats, tinted glass and a remote control driver's door mirror as standard. As usual it's a German car that comes at the bottom of the list when it comes to standard equipment, though the GTi has tinted glass, a laminated windscreen and a locking petrol cap. New for 1980, will be alloy wheels and a five speed gearbox as standard — needless to say the price will be more than at present.

Costs

It's almost as if the needle has got stuck ... once again the Volkswagen Golf comes out tops. It may have fuel injection but its consumption is so far ahead of its rivals it looks almost like an economy car. The lowest mpg we achieved was 23.0 mpg, coincidentally the same as the Government test figure for the urban cycle, while the best was over 30 mpg — with a fifth gear it will be even better. In contrast, of the others, only the Escort managed to scrape above 20 mpg, and the remaining pair shocked us with their incredible thirst.

Servicing costs are going to penalise the Vauxhall owner. His Chevette needs major attention every 6000 miles, while the VW and Talbot have 10,000 mile intervals and the Ford needs looking at every 12,000 miles. Parts costs are much the same, but woe-betide any Vauxhall owner who bashes a front wing, for nearly eight hours is the quoted time to put the matter right.

Insurance costs for such high performance cars are going to be high no matter who the driver, but we found quotes favoured the Sunbeam. Ease of servicing is an important consideration and although the Escort is usually sold by Ford Rallye Sport dealers, its common running gear will provide no surprises for the bulk of Ford's 1250 dealers and agents. The two other British cars have dealer representation around 640, while Volkswagen have 360 dealers. Warranty periods are the usual 12 month and unlimited mileage deals, though Volkswagen also have a six year warranty against rust.

Verdict

It became obvious early on in this test that for everyday use there was only one winner — and that was a winner by some margin. The Golf GTi is a truly remarkable car being quick yet frugal, sporty yet practical, and understated. It is not perfect — quite. We feel the interior lacks style and we know from experience it does not take long before it gets to look shabby. We also feel there is perhaps too much painted metal around inside for the price, and perhaps a little sound insulation under the bonnet would not go amiss either.

Runner-up must be the Escort which too, is quick and practical, but loses out on refinement, interior space and roadholding. In its favour it is well priced and lacks some of the mechanical complication of the Golf. The Vauxhall and Sunbeam are in a different class altogether. We suffered near accidents at traffic lights with both cars thanks to the fouling of the plugs — the cars will pull away from lights only to stutter and near die, causing heavy braking from behind. The only answer is to rev the engine high and drop the clutch as if doing a standing start at the test track, no wonder fuel consumption was high and looks from other drivers disdainful.

For our money, it's got to be the Golf all the way.

IMPRESSIONS: FORD ESCORT 1.3GT SPORT

The end of the year always has a few motoring manufacturers coming out with 'new' cars for the following year. This year has been no exception with first Opel and then Mazda naming their cars for the '80s. Ford, not to be outdone, have also uprated their current range of Escorts and come up with the 1980 Series Ford Escort.

The 1980 Series Ford Escorts were released in October 1979 and included further refinements and improved features at no extra cost. Most outstanding of the improvements are a new two-spoke padded steering wheel for all models except the 1.6 Sport and houndstooth material for the seats. Additional sound insulation has also been added to improve passenger comfort. In fact, sound insulation material is now up to Ghia standards. The headlining has also been changed to that of the Ghia.

Externally, the all black radiator grille carries the Ford oval at its centre. Only the Ghia still retains a chrome surround. To set off the grille are new rectangular driving lamps found on the Sport. The bumpers have rubber overriders similar to that of the 1.6 Sport and the 1.3GT.

A useful addition is the facility for intermittent wipe which complements the standard two-speed wipers.

So much for the generalities of the 1980 Series Ford Escorts.

The new addition to the Escort range is actually a dolled up version of the good old tried and tested 1.3GT. By adding bright colours and suitable macho striping the 1.3GT looks lovlier and brighter. A chin air-dam and bootlid spoiler ala the 1.6 Sport justifies calling it the 1.3GT Sport. Wide wheels (5J x 13) with that sculputured steel look and extra offset that increases track by 19mm. front and rear give the car a more sporty stance. The rest of the car is all Escort GT and those familiar with the marque would

find it no different from the other models in the range. The last time we tested an Escort 1.3GT was in early 1978 (see Auto International Vol. 2 No. 9) and the most recent test of an Escort was the one we did on the 1.3GL (see Auto International Vol. 3 No. 9). Most of our comments for the above two models hold true for the 1.3GT Sport as well, especially those concerning passenger comfort, boot space, ride and handling. So, we shall just skirt through those aspects as we go along and discuss the car more from the Sport point of view.

Firstly, the mechanicals. The engine remains the same as that on the old GT with a bore and stroke of 81mm amd 63 mm respectively to give a displacement of 1298cc. Contrary to other reports in the press the engine puts out a healthy 52.2kW at 5,500rpm instead of the 41.9 kW of the GL. Torque is average for the class at 95.2Nm at 4,000rpm. The OHV engine uses premium fuel because of its 9.2 to one compression ratio and is fed by a Weber twin choke unit. For a 1300

the engine puts out power on par with its contemporaries and even does better than some of the Japanese offerings.

The butter-smooth Ford gear' ox needs no introduction. Its slick movement makes it a joy to use and for the 1.3GT Sport the gear ratios are suitably stepped to provide smooth, brisk acceleration whilst remaining within the torque band. We still say that is one of the best around. The synchromesh is strong and there was no way we could beat it. The cable actuated clutch is light and positive in use and makes for smooth takeup of drive.

With most makers converting to trailing links and coil springs for their rear suspension setups Ford's semi-ellipitic leaf spring axle system still remains outstanding. In front, MacPherson struts keep the front wheels in contact with the road. To reduce the amount of body-roll there are antiroll bars front and rear. Spring/damper rates are well matched to give the car a taut ride with a reasonable amount of passenger comfort.

The GT goes Sporty

ESCORT 1.3 SPORT

The steering is our favourite rack and pinion system. With 3.5 turns lock to lock and a turning circle of just under 9 meters the Escort is quite manoeuverable. So for the little woman who wants a dressed up GT, the 1.3GT Sport would be worth considering. Steering effort is light.

The disc/drum brakes with vaccum servo assistance are more than adequate for the speeds the car is likely to reach. The feel at the pedal is progressive and feedback is good so that the wheels can be kept at the verge of locking during hard braking. Servo assistance is just right — it assists the pressure applied at the pedal instead of multiplying it like on some other cars. The brakes remain fade free during normal use, including fast inter-town driving. Even taking the car to the extreme at the race track did not bother the brakes.

Inside the car the new two-spoke steering wheel was the first item that caught our attention. Next were the houndstooth patterned cloth covers for the seats. The seats provided good sup-

port and located the occupants quite well even during exhuberent driving. We missed the head restraints. What we didn't like was the way the seat belt buckles were located. A tall driver with the belt correctly adjusted finds the buckle dangerously close to the lower stomach. The stalks are too long.

The instrument panel is the standard Escort GT unit and comes complete with tachometer. The view of the instruments has been improved with the new steering wheel. Its two spokes stant downwards from the boss thus giving more view area through the wheel. The dash remains the same with eyeball vents at each end.

If adding the Sport motiff to the GT is expected to transform the car then many prospective owners are going to be disappointed. What the Sport does add is the macho looks and aerodynamic aids. But overall feel is that the car is still the same mildly powered Escort which is so much better in 1.6 guise, whether in Ghia or Sport trim.

Our car was air conditioned but the 1.3 peppiness was still noticible. The car is not a street-rod, it just looks the part. Perhaps it is this feature which attracts a few followers. Women drivers will love the smart bright colours and the light controls. Driving in city traffic is easy, adequately comfortable and quite relaxing with the aircon on. Visibility is good and the shorter drivers will appreciate the highish sitting position which permits a good view of the front extremities of the car.

The car remains a safe understeerer under normal driving conditions but will let go at the tail when provoked. The precise steering allows the slide to be easily caught and exhuberent drivers will find the 1.3GT Sport a fun car except for the lack of real power. A couple of laps at the track soon revealed that the brakes and suspension were well suited to the power and weight of the car. Even the wet track was no real hassle to the car. It slid with plenty of warning and there was time to apply corrective

Another item in the Sport package is the boot
lid spoiler and bumper overriders.

The airdam, auxiliary lights and sports stripes
provide dash and style to what is basically 1300GT

The sporty two-spoke steering wheel and
houndstooth fabric trim are new on the 1300GT Sport
and help give an average car a bit of panache.

measures to provide smooth driving. On the open road the car felt skitterish, especially when it was undulating as well as wet but this was the top of the car's performance and sedate drivers will not be able to fault the car's roadholding and handling.

The performance figures speak for themselves with the Sport showing slightly better acceleration times than on our previous test 1.3GT. The slight increase in front and rear tracks is hard to assess but if Ford think that it is an improvement than we are sure that it works although it is not noticible.

For the price, the 1.3GT Sport is a reasonably good buy. For the same price as the 1.3GT ($15,432.24 on-the-road) the Sport comes with the extra lights, spoilers and bright paint and the good old Ford dependability and easy availability of parts. Besides, practically any roadside mechanic worth of his shack is capable of working on the straightforward mechanicals. •

SPECIFICATIONS

MAKE:	Ford
MODEL:	Escort 1.3GT Sport
PRICE:	$14,881.04
IMPORTER:	Ford Malaysia

ENGINE

Capacity:	1,298cc
Compression ratio:	9.2:1
Carburettor:	1, Weber 32DGV
Maximum power:	52.2kW at 5.500rpm
Maximum torque:	95.2Nm at 4,000rpm

TRANSMISSION

Gearbox:	4-speed, fully synchronised
Clutch:	Sdp, diaphragm, cable actuated
Ratios: 1st	3.337
2nd	
2nd	1.995
3rd	1.418
4th	1.000
Rev	3.876
Final drive	4.125

CHASSIS & RUNNING GEAR

Construction:	Integral
Front suspension	MacPherson struts, coil springs, anti-roll bar.
Rear suspension:	Rigid axle, semi-elliptic leaf springs, anti-roll bar.
Steering:	Rack and pinion

DIMENSIONS (mm)

Wheelbase:	2.400
Front track:	1.276
Rear track:	1.340
Overall length:	3.983
Overall width:	1.570
Overall height:	1.384
Ground clearance:	125
Kerb weight:	860kg

ELECTRICS

Headlamps, number & type:	2, rectangular, halogen
Battery:	12 volts, 38 Ah
Alternator:	12 volts 420 Watts

WHEELS & TYRES

Wheel size and type:	5J x 13, sculptured steel
Tyre size & type:	155 - 13, 80 series radial

PERFORMANCE

Maximum speeds in KPH:	1st	50
	2nd	84
	3rd	120
	4th	151

ACCELERATION (in seconds)
Through the gears

0 - 50kph:	4.5
0 - 80kph:	9.6
0 - 100kph:	15.9
0 - 120kph:	23.6

YJS·507

PRIDE, WITH A CERTAIN AMOUNT OF PREJUDICE

When his aging Datsun 1600 began showing advanced signs of decomposition, Canberra motoring writer Paul Gover set out to find a replacement. He chose an RS 2000 Escort, which he is learning to live with . . .

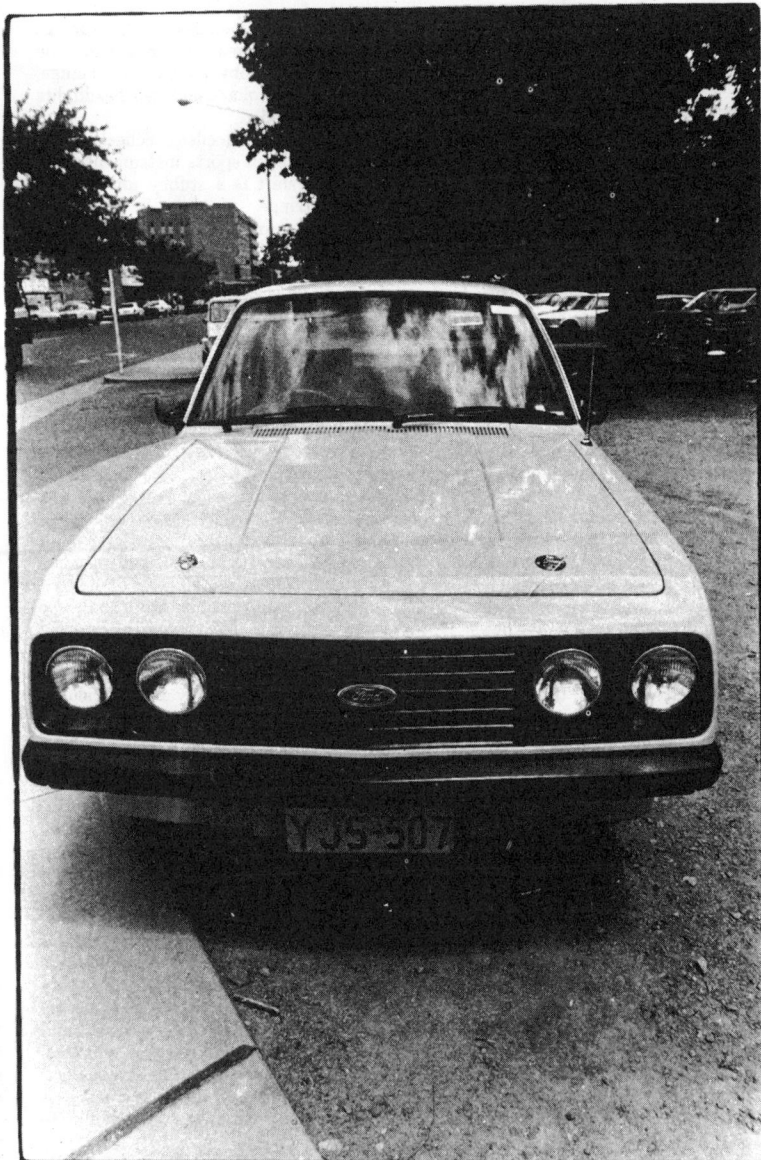

THE PLEASURE and challenge of attacking a winding road in a sporty sedan is being eroded on all sides by forces bent on seeing that enjoyment taken from the motorist of modest means.

Speed limits, the call for fuel economy, escalating new-car prices and many other factors are resulting in cars that are merely transportation.

Such social and economic pressures have already led to the downfall of the V8 super-car and the true sports car — indeed, even the modified van — is coming under seige!

The trend today is toward compact two-litre fours and the *good* news is that manufacturers are recognising the demand for such cars, ones that also give motoring an enjoyable feel once more.

Toyota has responded with what it has christened its "specialty cars": the Celica and T-18. Unfortunately, while these models are well equipped, they also continue the Toyota tradition of producing cars better suited to boulevard promenading rather than hard motoring.

Datsun's offering in the same area is the 200B SX, a tarted-up version of the 200B. It has a number of good points, but suffers from most of the 200B failings.

Which brings us to Ford's RS 2000, the most promising of the new breed of sporty but sensible sedans — without spending more than $10,000, there are few cars with the sporty pretensions of the newest Escort, but it does have its problems.

While the *RS* tag has unfortunate connotations for an Australian car, the label has long been used in Europe to signify the sports versions of Ford's otherwise-humble Escort. In much the same way that GT Falcons, Monaros and the various high-performance Toranas appealed to the Australian enthusiasts, the numerous RS Escorts were virtually assured of sales success in Europe, courtesy of the many race and rally wins achieved by the *real* RS cars,

The most outrageous of the RS breed were the RS 1600 and RS 1800 models. They were powered by twin-overhead camshaft engines driving through five-speed gearboxes

to heavily modified rear ends, as Ford played the same homologation games which led, in Australia, to the "Bathurst Special" Holdens and Fords.

The RS 2000 was added to the Mark I Escort range as a less-specialised sports sedan and was carried through into the Mark II series. It was a small batch of this model which in 1967 first introduced the RS 2000 to Australia, and those not snapped up for racing or rallying quickly found their way into enthusiasts' garages despite their $8500 pricetag.

RS Escorts next came to prominence in Australia in 1977 when Ford formed its rally team, with Colin Bond and Greg Carr campaigning the RS 2000 before moving on to the more-exotic RS 1800. The team's success paved the way for the re-introducion to Australia of the off-the-showroom-floor RS 2000 last year.

Its re-introduction also coincided with a crisis in my motoring history — my faithful, but heavily-modified Datsun 1600 shed an engine.

PRIDE, WITH A CERTAIN AMOUNT OF PREJUDICE

As this looked like the first of a series of failures which would eventually prove terminal, the reluctant decision was taken that the 1600 would have to go. Choosing a replacement was a harder decision . . .

Having driven a series of heavily-modified Datsun 1600s, including a rally car, and many test cars, there was little on the market that seemed a likely successor.

My buying criteria were performance to match a 180B engine and the gas suspension and Pirelli tyres of my now-defunct 1600, together with ergonomics to equal the Recaro seats, sports instruments and Momo wheel of the old car.

As a motoring writer for a Canberra newspaper, auditioning likely successors was easier (but more involved) for me than for most motorists and that's how the RS 2000 entered my life — by way of a pre-release drive of the four-door version for the newspaper.

After much racking of conscience and re-reading of overseas test reports, the decision boiled down to either a second-hand Alfa or the Escort, and it was probably the prospect of owning my first new car which tipped the scales.

At $6767 on the road, it seemed to fit most of my criteria, which also included carrying two adults and two children, looking a little different, returning enjoyable yet economical motoring — and being a sound investment.

Having made my choice, it was merely a case of waiting six weeks for a yellow, two-door, laminated-screen RS 2000 to be delivered. I fared better than some, for the waiting list grew to three months once the car hit the market.

There were a number of things that set the RS 2000 apart from lesser two-litre Escorts, but all contributed to its driving pleasure. The suspension, for example, is that fitted as the sports handling pack to other Escorts, and includes firmer springs and shocks and a rear anti-roll bar. The plastic snout is the obvious exterior change, incorporating four quartz-halogen headlights and deep chin spoiler.

Inside, there are excellent Scheel sports bucket seats and full sports instrumentation, while the gearshift is a stubby unit with a tight *sports* change.

The running gear is the same as all other two-litre Escorts, though the RS has a slightly better top speed, thanks to its aerodynamically cleaner shape.

Fuel consumption for my car has been about 8.5 km/l (24 mpg) around town and just over 10.6 km/l (30 mpg) on a trip.

Performance has proven very good for a car of its type, and, though it is no electrifying road burner, it will cruise happily well over the posted speed limit if I'm feeling that way inclined.

Kicking open the scond barrel of the downdraft Weber carburettor gives a pleasant burst of induction roar, reminiscent of the twin 45 sidedrafts on my fondly-remembered Datsun.

Though the gearbox ratios are a little too widely spaced to be ideal, working up and down through the box is a delight once the short and notchy shift is mastered. But it can be a problem finding reverse, particularly when the gearbox is cold.

The handling of the RS is very good, even if the extra weight in the snout causes it to understeer more than the Escorts fitted with just the sports-handling package. However, the car can be flung around with gay abandon, thanks to the pin-sharp rack and pinion steering, and is equally at home in traffic or on the open road.

With my leaning towards rallying, and my style of driving somewhat influenced by this sport, the car has been tried thoroughly over dirt roads. Here the handling is again very good, but it does highlight the RS's one big flaw: the rear end is extremely skittish, and is prone to jumping sideways on corrugations or potholes on both bitumen and dirt. It's perhaps significant that the European RS 2000 is fitted with twin trailing links to locate the rear axle . . .

After an initial braking problem fixed by fitting a new master cylinder, the brakes have proved faultless, though the pedal pressure needed for maximum stopping is high.

The other big disappointment with the car is also the bugbear of many Australian cars: the finish. The paintwork on my RS displays runs and overspray, particularly where the matt-black paint-outs have been applied over the main color. In addition, the snout is poorly fitted. Inside, things are much better, though dash rattles are beginning to develop after 10,000 km.

On the plus side, there is plenty of room for two adults, two kids and their attendant toys and paraphernalia, and the boot is particularly big for a car the Escort's size, since the fuel tank is located below the boot.

For night driving, the headlights are good, though they have been vastly improved by fitting 100-watt globes to the high beams.

The only other addition to my car has been the sound system, featuring a 25-watt-per-channel amplifier coupled to a radio-cassette and speakers.

The next objective is to improve performance and economy, and to give a little attention to the suspension, tyres and gas shocks, which should see the car handling even better. After that, who knows?

It's still a long way from becoming as close to me as my old Datsun, but one day . . .
*

AUTOCAR, w/e 12 June 1976

RS2000 weekend

CONTINUED FROM PAGE 39

side to keep the inside wheel down, my halogens on full beam trying to out-stare the moon now straight ahead between the banks of the lane. Good marshalling helped, and the blessed fact that it was dry. It was a new experience, hill-climbing at night, and more exciting. We blessed the sight of Observed Section Ends, but had to stay in first gear for the rest of the long climb continuing after. Tyres were pumped up again, one pump per back wheel, on a moonlit slope just after five, a calm and beautiful place after the anxious hill.

Dunster church was a bluff silhouette against a hint of light 20 minutes late, and the climb of Porlock heralded the first time I have begun an Exmoor crossing in dry weather, though out to sea from Countisbury it was hazy.

Beggar's Roost was next, and I have firm evidence that the Beggar reads *Autocar*. Or at any rate the issue of 10 April, when I wrote blithely "in my small knowledge, it has not been the worrying thing I had expected," adding, without wishing to appear condescending at all, that "I would be interested to hear from older competitors who know the Roost if it has got easier in recent years?" The first sign was when two of the three Imps in front of us reappeared at the start line backing down the slopes. There wasn't time to go and inspect the hill. We started all right, very well, and then came round the corner. There was a broad area of chewed-up loose stuff. We drove at it and made reasonable headway. But the car wasn't maintaining speed. It was slowing. It was still going though. It *was* slowing, the back tyres spinning more than they should be allowed to. There seemed no better route to take; I must plug on, but couldn't make myself back off the throttle, fearing it would stall. We stopped, wheels spinning after 15 yards from the line. Damn. The Beggar had won.

It was all the more annoying when with only a small push from marshals and spectators plus a more gentle right foot we re-started and finished the climb. I should have had more faith in the excellent wide spread of torque of the two-litre engine, as I did later on. It was comforting later on to learn that over 100 cars had failed on Beggar's Roost, which proved the stopper for the Trial this year, and that the gentleman in charge of it had got permission to prepare the way with his JCB. Before the last war, when the Roost first became less difficult, the local council, wishing for the sake of local business to keep the customers, had spread loosening stuff, so the tradition was being observed. It served me right of course.

After a compulsory fuel stop at Blackmore Gate at 6.13 came a dry Orange, a grassy rutted track, then Sutcombe, nervousness about which dispelled the peak of tiredness which had come upon me. The approach to it is along a narrow boggy track which raises doubts

before you start climbing. We stopped on some soft green stuff to let tyres down to 12 psi, and had doubts leaving it. But the climb (and the restart which this year I remembered) went all right. We were getting more and more confident of the RS2000, which with the same ungainly-looking extra long back shackles and a lot of ballast was proving superbly competent. The glorious thing about it was the way it didn't just climb, but accelerated up these often appalling-looking tracks. Darracott was the best thing of the day for the car and for its occupants, a bumpy, very steep track with some exhilarating, tightly wrapped narrow near-hairpins which one had to rush at, foot down for fear of stopping, yet equally frightened of understeering outwards to bury the nose in the bank. The 53 per cent static rear wheel loading plus the squodginess of the superb Cinturato M+S tyres at 12 psi avoided that however; you could swing it neatly, the car feeling virtually neutral overall (though oversteery in this state on a hard surface) when the understeer had been taken account of, so that it took these bends very tidily — and excitingly. The sideways-adjustable ballast — Father in the back seat — helped a lot too.

Crackington was a near thing, the driver ascending the last part which was curiously much worse than the rest with his mouth

Threlfall's Ford behind our one waiting to get to grips with Ruses Mill; passenger Barker just visible behind

hanging wider and wider open, trying to make himself back off the throttle. We did, *just* enough, the tyres cogging into the muck and lifting us through. That last part seemed to last a very long time, but suddenly we felt that precious grip returning and the engine still turning over, prompting us both to shout five yards before the line, grittingly, "We've made it." Lovely feeling; the tensest one so far. A marshal had suggested that it had been doctored a bit, and we agreed, declaring an active interest on the part of the locals to be a healthy thing. One of them at the top blandly wondered "where on earth you've picked up all that Cornish mud on thus zunny day?" He knew a thing or two.

Lewdown was some people's lunch, but our breakfast stop, at 10.47. Ruses Mill was next, a most picturesque spot, a former haunt of the MCC. It is steep and hard-surfaced, and the stop and re-start there nearly did for the Escort's clutch, not relieved as elsewhere by friendly wheelspin. We left smelling somewhat whilst the clutch recovered itself. The sun was gloriously hot, and we wished with all respect to the Ford that we were back in last year's Jensen-Healey. It was perfect open-car weather. However the Escort was doing us proud, and paradoxically for a car with excellent closed-window ventilation, it is extremely pleasant with the windows down, because there isn't too much side blast. The rest of the trial went — I was going to say smoothly, but you cannot use that word of any of the three remaining places, Brickannan, Newlyn Downs and especially Blue Hills Mine. Some-

A Dellow fellow competitor bowls by during a breather taken on A30

one had oiled the track at Brickannan they said, and it did look it; Newlyn Down, a set of timed tests in a quarry was great fun, the Escort's swingy ways suiting it perfectly, and Blue Hills was as ever brutal, banging and the superbly satisfying finale it always is. There are unpleasant ramps there, which remembering last year and the way the Healey's nose pitched so wildly, provided a very good demonstration of the difference between worn and good dampers. The RS 2000 romped up; I had to hold it back at times once over the worst to save hurting it. The Imp behind, Betson's 998 one, was not so lucky. He appeared, towed by a Land-Rover, and making a most expensive-sounding clanking noise as a drive shaft, relieved of its doughnut, flailed the body.

We left, helped out of the Blue Hills area as we had been into it by excellent policemen on point duty — there was a bigger crowd than I have seen before watching there. Newquay and the official finish at The Cornish Chough we reached before 4pm, claiming, with good humoured curses on that perfidious Beggar, a second class award. A most desired beer followed, and very interesting chat, then we were off again, east this time, leaving at around 4.30 for home with nothing more than our Finisher's Certificate and two pots of clotted cream for family appeasement.

The drive back went very well, with a meal stop halfway, the Escort's superb comfort and easy surplus power making light work of the 251 miles to London. The only

detail which tired one unnecessarily was the stiffness of the car's dipping stalk and the distance it is from the steering wheel. It was 10.40pm when we arrived, weary but very satisifed with the day's work.

I got up on Sunday at 9.30 to find my brother, bless him, washing the RS. Between us, we changed the shackles back to standard, and loaded a roof rack on which went two bunk beds which were wanted by my brother-in-law in Scotland. On starting up the engine I discovered the tank was virtually empty; the standard tank holds nine gallons only which like the silly little horn is inadequate for a car of this character and price — not that £2,983 is anything but excellent value by today's standards.

I didn't leave for the north until 4.25 pm, bidding farewell to my navigator who, like me, felt more than fresh enough to drive to Scotland. Naturally, to speed things as much as possible, I went M1/M6, A74. The damnably mean idiocy of continuing with the farcical 50/60 limits meant that motorway had to be a better way. It was hot but thundery. One of the pleasing things about cross-Britain travel in mid-Easter is that most people have done their motoring, and so traffic was light. 477 miles later those superb halogen lamps were piercing the midnight sky on the road to my bed, scattering sensible hares and mesmerizing silly rabbits. The BBC had relieved any loneliness with some excellent music and a performance of *Henry V*, and those incredible seats had prevented any sign of discomfort — they simply fit all points they touch and coupled with the qualities of the car made the journey so much easier.

Monday saw the RS way below us on the road by the bridge in the Sma' Glen as we walked, supposedly on a compass bearing, back to where we were staying. The hills were alive with grouse; hare changing their coats from winter white to summer brown, patched; curlew crying that curious up-turning note or chuckling to themselves with an even more wonderful bubbling song; lapwing by the score, and, we thought, a snipe. It was a way of using a car which I love — to drive to a place, and walk to another, knowing that someone can drive you back to collect the car afterwards. We foolishly didn't believe our bearing, so made a curving trudge through the heather, the townsman (me) having to stop many times whilst my fit brother-in-law waited patiently. The next morning was spent teaching the driver to hit a golf ball — with a golf club; he found it easier to hit it with a car — amid the squabbling of wild geese feeding in the rough beyond, then arising high above us, still talking to each other as they wheeled away. Then we returned reluctantly, thankful though that removing the roof rack and its load had improved the fuel consumption back by 10 per cent to 28 mpg, and for the loan of a highly satisfying Ford. I could live happily with an RS 2000 (though I'd remove the thing on the boot). □

In the club

*Look at your average special stage rally entry list. Still dominated by Fords, RS2000s and RS1800s, wouldn't you say? Finding out why, **CCC**'s fearless forest tester, Fred Henderson, has been hard at work comparing two of the best, both on the dyno and on the track*

The two vehicles in question this month are a Ford RS2000 (Group somewhere between One and Five) and a Ford RS1800 (definitely Group Four). The latter car is probably better known as the ex-Malcolm Wilson Rallysprint machine, seen rolling every Saturday on BBC's *Grandstand* programme, and is now owned by Newcastle garage owner, Brian Stanners, who is quick to point out that not much of his car is actually seen on the 'box'.

Brian first started rallying back in 1976 using a standard Ford RS1600, which he was later to modify and use with great success in the ANECC road rally series. Although in 1977 he won no less than 10 road rallies, the actual championship eluded him at the last hurdle. His present car was purchased for the 1980 stage rally season and is in continual use at the present time.

Our owner/driver of the RS2000 is Pete Tyson from Cumbria, a self-employed builder who competed in his first event in 1975 using a Mk2 Cortina. Pete has shown considerable promise this year, driving his present car, and finished a secure fourth on the Lakeland Stages, his best result to date. The car, like Brian's, was purchased from someone else. On this occasion, though, he built up the vehicle and not only ran out of money, but could not get it to run right either.

Just standing back and looking at the two vehicles, there really aren't that many differences; the Group Four car features a front air dam and 6in. (or more) Minilites whereas the smaller car has no air dam and 5½in. Revolution wheels. While these two items distinguish the two cars and other areas remain the same just looking at the vehicles, it is easy to sense a full Group Four car and a Clubmans hobby; it is that same feeling that separates a works Fiat from even the most carefully constructed replica.

While Brian's car uses the well proven 2-litre BDA, Pete Tyson's car uses a converted Pinto engine, or does it? The RS2000-type engine has been extensively modified by the Ipswich tuning firm of Holbay. Firstly in order to aid cooling and avoid possible air pockets, something which has caused more than one head gas-

ket failure on a tuned SOHC engine in the past, the cooling system has been completely reworked as follows: Instead of the water passing along the cylinder head and out of the thermostat housing, it is released through the top of the cam cover.

This is achieved by removing the two plugs under the camshaft and effectively piping water up through a pair of tubes to the top of the cam cover. An alloy housing (2-into-1) then takes the water via a thermostat to the top of the radiator.

For its return journey to the engine, water leaves the bottom hose and flows into a separate

> **In all my years of driving in many different types of vehicles, I have never made such a pig's ear of getting away from a standing start as with this car. I would go as far as to say that out of some 50 starts, only about two were made without some sort of jump or stall.**

water pump which is bolted onto the normal alternator position. This pump (which looks like a V4 Transit type) then pushes coolant back into the block via core plug holes on the side of the block.

What this means is that the existing thermostat housing and

water pump "hole" are blocked off and the alternator is rather awkwardly mounted upside down on the manifold side of the engine, with its diode mounted on the inner wing.

A Group One type cylinder head is used with Group One valves, activated by a camshaft described as a 509M. The lower half of the engine is really beefed-up with the use not only of steel crank and rods, but Cosworth pistons described as BDA-type.

As the engine is dry sumped, a different oil pump is used and this is driven with a separate toothed belt, as is the water pump; while the alternator is driven by a more conventional form of belt. Carburation is by a pair of 45 Weber units and a four-into-one exhaust manifold is used.

Besides the engines, the similarities again appear. Both cars use the same type of twin plate clutch, although Stanners' is hyd-

of a left foot rest for the driver (in both cases). Surprisingly, even the RS2000 used 24 volt starting, with the batteries encased in neat and tidy boxes in the boot on the opposite side from the dry sump tanks. The only difference in the boot area concerns the fuel tanks. Whereas the Group Four Escort uses a 18½-gallon foam-filled bag tank, the RS2000 has to make do with an ex-Capri 11-gallon tank. One slight difference that is not apparent to the eye is the

Misfire prevents revs being used

DRAINAGE PIPES

RPM x 100

——— BDG 2000 cc (RS 1800)

- - - - Holbay Ford 2000cc SOHC

raulically operated whereas Tyson's is by cable (more on this later!). Four speed gearboxes are used on both vehicles, both with Quaife innards, although the RS1800 uses a slightly lower first gear, supposedly specially developed for a quick getaway.

The rear axle on the Holbay car is a narrow Atlas (ex-Capri) with normal drum brakes and a ratio of 4.6 to 1, while the Group Four car has the normal slightly wider axle, 5.1 to 1, which also incorporates fully floating hubs and solid 9½in. disc brakes. Both axles are located by four equal length links, slipper springs and Panhard rod. Rear springs are rated at 145lb on both cars.

Similarly, at the front, both competitors use 190lb springs on Bilstein struts, although the Group Four car has adjustable springseats whereas in the RS2000, they are fixed. Ten inch vented discs with AP four pot calipers provide the stopping equipment on the front of the more powerful car, while the lesser output of the Holbay machine means that standard RS2000 Group One front discs do the trick, the latter not being equipped with any form of brake balance adjustment, just a tanden master cylinder. Naturally a World Cup crossmember and high ratio rack are common to both cars.

As mentioned earlier, different road wheels are used, although both drivers swear by Dunlop tyres.

Naturally, both bodyshells are strenghtened by the addition of 14-point roll cages and various little gussets. I liked the presence

presence of glassfibre wing extensions on Tyson's car, while Stanners' vehicle has aluminium extensions. What the financial implications of this are, I am not sure.

Before the cars were subjected to having me strapped behind the wheel, they were taken to McDonald Racing to have their respective power outputs checked on a rolling road. This is the equipment we have used for all our performance testing with the CCC Chevette, and while the figures

> **Generally the car handles so well that the brakes are not needed as much as on some other cars I have driven, but it remains a shame that something so intrinsically good should be so hampered in this way.**

produced may or may not be accurate, the comparison is beyond dispute. In fact the equipment seemed to me to be fairly accurate.

The vehicles are placed in rollers with the front wheels firmly chocked. The car is then run up in third gear. At various points, as determined by the operator, the electronic retarder is engaged and the bhp figure recorded. In real terms, the tractive effort at the wheels is being measured.

And so to the forest. . . . Sitting in the RS2000 is an easy and relaxing occupation with the pedals (standard in appearance)

The power just keeps on coming and coming in a great adrenalin rush and the vast rev range means one can just drive between the corners using the throttle and holding onto gears.

in the club

being pleasantly 'to foot.' The steering wheel felt well placed, or should I say, the seat is in the right position, not too low down. The engine started instantly with a turn of the key, although it was easy to sense a highly tuned engine, perhaps by the vibration that seemed to accompany the World Cup-type engine mountings.

Now the fun begins. Pressing the clutch either required bionic assistance or at least two feet. Since neither of these things are practical, the pedal has to be pressed as best one can. The gear shift, however, is just magic with very little movement between the ratios, but at the same time it is easy to detect which gear the lever is actually in.

In all my years of driving many different types of vehicles, I have never made such a pig's ear of getting away from a standing start as with this car. I would go as far as to say that out of some 50 starts, only about two were made without some sort of jump or stall, the clutch was that desperately fierce and heavy (although I would add that Tyson just drove it like an ordinary car!). Once on the move, (let's forget about the clutch), the gearshift and gear ratios were a delight, and this combined with the smooth power curve (see chart) of the engine meant that the car moved along impressively.

The power unit seemed really torquey and I saw no point in exceeding 7000rpm. The misfire that was present on the rolling road didn't show up when actually driving the car in anger. The one thing that really struck me

about driving an Escort is the turn-in ability; something that I have not experienced in any other make of car. Probably the best way to describe this is the fact that the first third of a corner could always be taken with the steering wheel as opposed to flicking the car sideways. This meant that really fast corners could be steered through with maximum confidence and less physical or mental effort.

Braking, on the other hand, I would describe as poor. At least 70 percent of the effort was on the front wheels. How Tyson drives like this, I don't know. It must really do credit to his ability, as the slightest application of the pedal on anything but a straight road sent the front of the car skittering to the outside of the corner. Generally the car handles so well that the brakes are not needed as much as on some other cars I have driven, but it remains a shame that something intrinsically so good should be hampered in this way.

It may or may not seem strange, but I had not driven a full Group Four Escort until I sat behind the wheel of Stanners' car. As mentioned earlier, despite some superficial similarity, it felt far away indeed from Tyson's car. I really felt I was sitting in a powerful machine, even before starting the engine. The clutch pedal, being hydraulically operated, had none of the problems of the earlier car, and was in fact a pleasure to use.

Once the engine burst into life and I set off down the stage, it was easy to see why so many people use these cars so successfully. The power just keeps on coming and coming, and the vast rev range (to 9000 or more if required) means one can just drive between the corners using the throttle and holding onto gears. Although I don't have any torque figures, they are obviously impressive as no 'caminess' is present, just a further boost of power at about 6000rpm. Handling and road holding are as good if not better than the smaller car, although in this case the brakes are incredibly strong and reassuring. Even on very slippery surfaces the car just digs in and *stops*. The only difficulty I encountered was in finding the correct gear for a given situation, as two gears only would do the job, mainly because of the vast rev range.

So there we have two cars – one priced at £10,000 and one at probably less than half that: in many ways, two very similar cars, but given a set section of special stage and the same driver, it would probably mean that the asking price, in terms of results, is something like £2000-per-second-per-mile. ∎

Power Figures	Tyson bhp	Stanners bhp
3000	60	68
3500	70	72
4000	82	100
4500	104	112
5000	122	132
5500	130	156
6000	140	170
6500		182
7000		195
7500		202

RS2000

Owner	Pete Tyson 19.10.51
Occupation	Builder
Sponsor	Car Bench
Origin of car	Purchase from friend
Model	Ford Escort Mk2 saloon
Engine	Holbay 2000cc SOHC
BHP (flwheel)	200 approx. 7000rpm
Ignition	Lucas (points)
Carburettors	45 DCOE Weber
Clutch	Twin plate (paddle), cable operated
Gearbox	4-speed Quaife RS casing
Axle	Narrow Atlas 4.6–1
Susp. rear	Slipper springs located by 4 links. Panhard rod
Susp. front	MacPherson strut standard type anti-roll bar
Springs	5 leaf 145lb/190lb
Shock absorbers	Rear Bilstein turret – Bilstein struts fixed
Brakes front	RS2000 G1 solid discs
Brakes rear	Capri drums
Brakes master/cylinder	Tandem – no adjustment
Wheels	5½in. Revolution
Tyres	Dunlop 175 x 13 MS
Electrics	24v starting, 12v System
Lubrication	Dry Sump tank in boot

RS1800 Group Four

Owner	Brian Stanners 20.11.46
Occupation	Garage proprietor
Sponsor	Nor-lex Land Drainage
Origin of car	Ex-Malcolm Wilson Rallysprint vehicle
Model	Ford Escort RS1800
Engine	BDG 1996cc
BHP (flywheel)	235 9000rpm
Ignition	Lucas Opus (Elec)
Carburettors	45 DCOE Weber
Clutch	Twin plate (Paddle) Hydraulic
Gearbox	4 speed Quaife RS casing, low 1st gear
Axle	Atlas 5.1–1 F/F hubs
Susp. rear	Slipper springs located by 4 links. Panhard Rod.
Susp. front	MacPherson strut standard type anti-roll bar.
Springs	5 leaf 145lb/190lb
Shock absorbers	Rear Bilstein turrets – Bilstein Strut adjustable platform
Brakes front	10in. vented discs, 4 pot AP caliper
Brakes rear	9½in. solid discs twin calipers, one for handbrake
Brake master cylinder	Dash-adjustable twin pedal box, 0.625 cylinders
Wheels	6in. Minilites
Tyres	Dunlop 175 x 13 MS
Electronics	24v starting, 12v System
Lubrication	Dry sump (tank in boot)

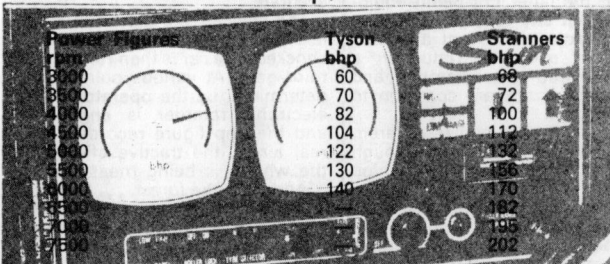

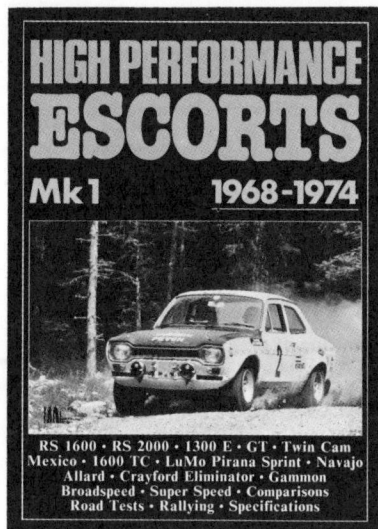

HIGH PERFORMANCE ESCORTS Mk 1 1968-1974

RS 1600 · RS 2000 · 1300 E · GT · Twin Cam Mexico · 1600 TC · LuMo Pirana Sprint · Navajo Allard · Crayford Eliminator · Gammon Broadspeed · Super Speed · Comparisons Road Tests · Rallying · Specifications

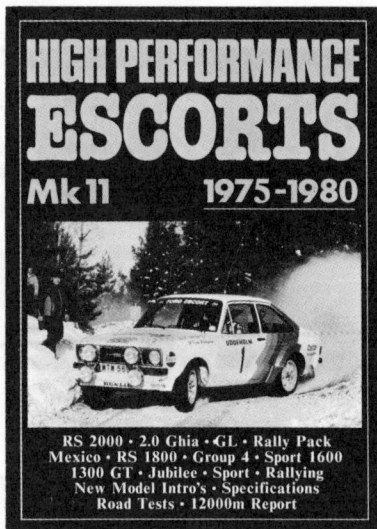

HIGH PERFORMANCE ESCORTS Mk 11 1975-1980

RS 2000 · 2.0 Ghia · GL · Rally Pack Mexico · RS 1800 · Group 4 · Sport 1600 1300 GT · Jubilee · Sport · Rallying New Model Intro's · Specifications Road Tests · 12000m Report

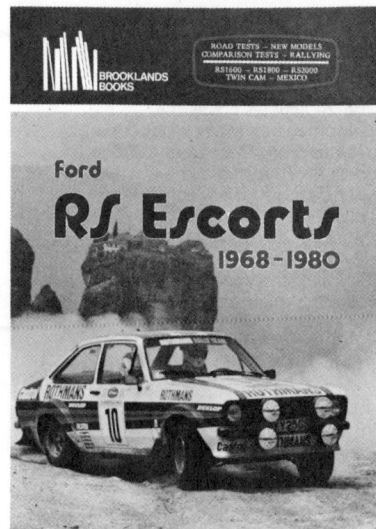

FORD RS Escorts 1968-1980

ROAD TESTS – NEW MODELS COMPARISON TESTS – RALLYING RS1600 – RS1800 – RS2000 TWIN CAM – MEXICO

HIGH PERFORMANCE ESCORTS Mk I 1968-1974

A total of 13 road tests are included in the 38 articles that retell the story of the more powerful Mark I Escorts from their introduction in 1968. High Performance models covered are the RS 1600 & 2000, the 1300 & 1600 Twin Cam, the Mexico and GTs plus versions from LuMo, Gammon, Allard, Navajo, Broadspeed and a 3-litre from Crayford. There are two comparison tests and articles on rallying and history.
100 Large Pages.

HIGH PERFORMANCE ESCORTS Mk II 1975-1980

The views of Britain, Australia and Malaysia are represented in the 25 stories that make up this book on the more powerful Mk II Escorts. Included are 6 road tests, a comparison test, a 12,000-mile report and rallying information. Models covered are the RS 1800 & 2000, Sport 1600, Mexico, 1·3 & 2-litre Ghia and GL, and the Jubilee 1·3 GT.
100 Large Pages.

FORD RS ESCORTS 1968-1980

Rally Sport Models covered in this book are the Twin Cam, RS 1600, RS 1800, RS 2000 and the Mexico. Some 30 articles trace the development of the RS models from 1968. They include 14 Road Tests, two Comparison Tests vs. the Viva GT and the Fiat 131 Sport, driving impressions, rally reports, touring, new model introductions and full specifications.
100 Large Pages.

These soft-bound volumes in the 'Brooklands Books' series consist of reprints of original road test reports and other stories that appeared in leading motoring journals during the periods concerned. Fully illustrated with photographs and cut-away drawings, the articles contain road impressions, performance figures, specifications, etc. NONE OF THE ARTICLES APPEARS IN MORE THAN ONE BOOK. Sources include Autocar, Autosport, Car, Cars & Car Conversions, Car & Driver, Car Craft, Classic & Sportscar, Modern Motor, Motor, Motor Manual, Motor Racing, Motor Sport, Practical Classics, Road Test, Road & Track, Sports Car Graphic, Sports Car World and Wheels.

From specialist booksellers or, in case of difficulty, direct from the distributors:
BROOKLANDS BOOK DISTRIBUTION, 'HOLMERISE', SEVEN HILLS ROAD,
COBHAM, SURREY KT11 1ES, ENGLAND. Telephone: Cobham (09326) 5051
MOTORBOOKS INTERNATIONAL, OSCEOLA, WISCONSIN 54020, USA.
Telephone: 715 294 3345 & 800 826 6600

MUSTANG MUSCLE CARS 1967-1971

Some 24 articles cover the development of the performance Mustangs from December 1966. Models dealt with are the Fastback 2+2 and coupé, the Mach 1 and Grande plus Boss variations. Engines reported on are the 302, 351, 427, 428, 429 and Cobra Jet 390. Fourteen road tests, a rod test and an engine analysis are included.
   100 Large Pages.

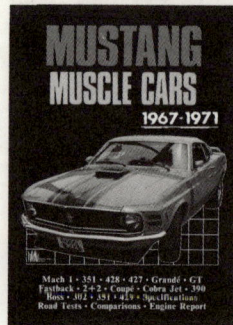

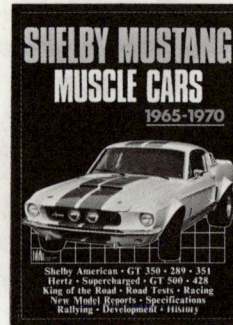

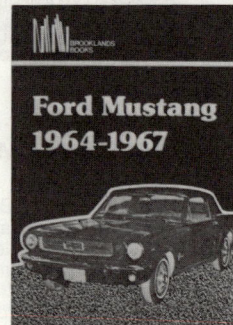

SHELBY MUSTANG MUSCLE CARS 1965-1970

Carroll Shelby's performance Mustangs are followed through 26 stories drawn from the US, Australia and Britain. They include 11 road tests, new model reports, full specifications, and stories on rallying, racing and their history. They cover the GT350, with both 289 and 351 engines, the Hertz and a Supercharged 350 the GT500 and the 'King of the Road' Model.
   100 Large Pages.

FORD MUSTANG 1967-1973

A total of 22 articles continue the Mustang story through to 1973. They cover 8 road tests, a track test, history, specifications, an owner survey and comparisons with the Camaro, Javelin, Barracuda, SS454 Chevelle, Duster 340 and the Shelby AC Cobra. Models covered include the Mach 1, Boss 302, the convertible GT, the Shelby and the Trans-Am and deal with the following engines — the 289, 302, 351, 390, 427 and 428.
   100 Large Pages.

FORD MUSTANG 1964-1967

Road tests, new model introductions, road research reports, comparisons with the Barracuda and Corvair, supercharging, touring, a drivers report and an article on a Wankel powered Mustang make up this book. Models covered include the 6 and 8 cyl. convertibles and sedans, the 350 GT & 390 GT, Ruddspeed and Bertone versions, the Shelby American GT 350 and GT 500 together with 4 articles on the preproduction prototype of 1962.
   100 Large Pages.

These soft-bound volumes in the 'Brooklands Books' series consist of reprints of original road test reports and other stories that appeared in leading motoring journals during the periods concerned. Fully illustrated with photographs and cut-away drawings, the articles contain road impressions, performance figures, specifications, etc. <u>NONE OF THE ARTICLES APPEARS IN MORE THAN ONE BOOK.</u> Sources include Autocar, Autosport, Car, Cars & Car Conversions, Car & Driver, Car Craft, Classic & Sportscar, Modern Motor, Motor, Motor Manual, Motor Racing, Motor Sport, Practical Classics, Road Test, Road & Track, Sports Car Graphic, Sports Car World and Wheels.

From specialist booksellers or, in case of difficulty, direct from the distributors:
BROOKLANDS BOOK DISTRIBUTION, 'HOLMERISE', SEVEN HILLS ROAD, COBHAM, SURREY KT11 1ES, ENGLAND. Telephone: Cobham (09326) 5051
MOTORBOOKS INTERNATIONAL, OSCEOLA, WISCONSIN 54020, USA
Telephone: 715 294 3345 & 800 826 6600

MINI MUSCLE CARS 1961-1979

A total of 34 articles trace the development of the powerful Cooper and 1275 GT Mini's from their introduction in 1961. There are 10 Road Tests, 4 Track Tests, a comparison between the Cooper S and 1275 GT, plus articles on racing, the Monte Carlo rally, history and a trip across the Sahara. Reports cover special tuning by Alexander, Speedwell, Oselli, Janspeed, Marcos, Broadspeed, Minisprint and Downton.
100 Large Pages.

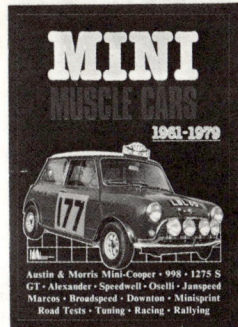

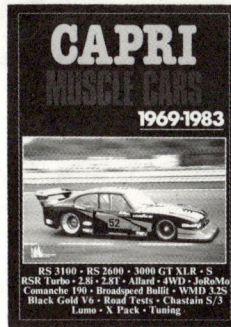

CARPI MUSLCE CARS 1969 - 1983

Some 30 articles retell the story of the most powerful Capris. Included are 10 road tests, a performance test, a 12,000 m. report, new model introductions and specifications. Models reported on are the RS 2600, RS 3100, 3000 GT XLR and S, the RSR Turbo, the Black Gold V6, the 2.8i, Tickfords 2.8T, the Comanche 190, Bullit, Chastain S/3, JoMoRo, plus cars developed by WM, Allard and Lumo. The 4 WD Rallycross 250 bph V6 is covered.
100 Large Pages.

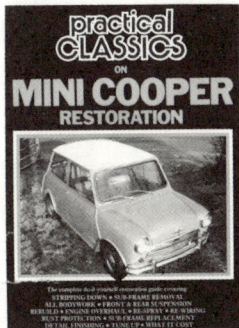

PRACTICAL CLASSICS ON MINI COOPER RESTORATION

Articles exclusively from Practical Classics' magazine describing in great detail the complete restoration of a Mini Cooper. Aspects covered include: stripping down, sub-frame removal, all bodywork, front & rear suspension rebuild, engine overhaul, respray, re-wiring, rust protection, sub-frame replacement, details finishing, tune up, what it cost. Fully illustrated with approx. 120 photographs.
72 Large Pages.

MINI-COOPER 1961-1971

Articles drawn from Britain, Australia, Canada, Ireland and the US make up the 37 stories in this book. Of these 12 are Road Tests, 2 Comparison Tests also a used car test, touring topics, plus new model introductions and specifications etc. are included. Vehicles covered are the Austin and Morris Mini-Cooper, including the "S" and engines of 997 cc, 1071 cc, 1275 cc and 1390 cc are dealt with. Specially tuned cars by Speedwell, Downton, Alexander, Taurus and Derrington are also reviewed.
100 Large Pages.

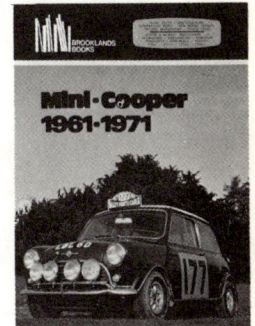

FORD RS ESCORTS 1963 - 1980

Rallye Sport Models covered in this book are the Twin Cam, RS 1600, RS 1800, RS 2000 and the Mexico. Some 30 articles trace the development of the RS models from 1968. They include 14 Road Tests, two Comparison Tests vs. the Viva GT and the Fiat 131 Sport, driving impressions, rally reports, touring, new model introductions and full specifications.
100 Large Pages.

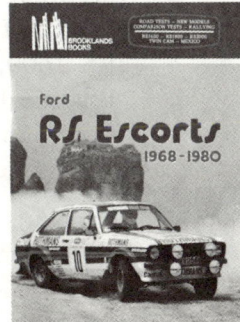

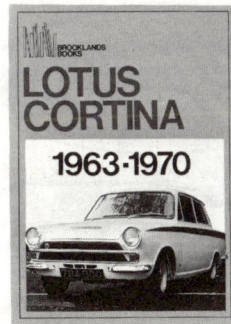

LOTUS CORINTA 1963 - 1970

Stories from Britain, Australia and the U.S. make up the 33 articles that trace the development of the Lotus Cortina between 1963 and 1970 There are 11 road tests, a comparison test, a road report, plus articles on development, racing, rallying, touring, model introductions and testing. Vehicles covered include the Mk. I and Mk. II plus cars modified by IWR, Westune, Taurus, Willment and the works.
100 Large Pages.

These soft-bound volumes in the 'Brooklands Books' series consist of reprints of original road test reports and other stories that appeared in leading motoring journals during the periods concerned. Fully illustrated with photographs and cut-away drawings, the articles contain road impressions, performance figures, specifications, etc. NONE OF THE ARTICLES APPEARS IN MORE THAN ONE BOOK. Sources include Autocar, Autosport, Car, Cars & Car Conversions, Car & Driver, Car Craft, Classic & Sportscar, Modern Motor, Motor, Motor Manual, Motor Racing, Motor Sport, Practical Classics, Road Test, Road & Track, Sports Car Graphic, Sports Car World and Wheels.

From specialist booksellers or, in case of difficulty, direct from the distributors:
BROOKLANDS BOOK DISTRIBUTION, 'HOLMERISE', SEVEN HILLS ROAD,
COBHAM, SURREY KT11 1ES, ENGLAND. Telephone: Cobham (09326) 5051
MOTORBOOKS INTERNATIONAL, OSCEOLA, WISCONSIN 54020, USA.
Telephone: 715 294 3345 & 800 826 6600